DJ.M 10-11

Collins
Revision

NEW GCSE SCIENCE

Science and Additional Science

Foundation

For OCR Gateway B

Authors: **Colin Bell**
Brian Cowie
Ann Daniels
Maureen Elliot
Ian Honeysett
Chris Sherry

Revision Guide +
Exam Practice Workbook

Contents

Fitness and health

Blood pressure

G–E

- Blood in **arteries** is put under pressure by contraction of the heart muscle so that it can reach all parts of the body.

D–C

- **Blood pressure** is measured in millimetres of mercury. This is written as mmHg.
- Blood pressure has two measurements: **systolic pressure** is the maximum pressure the heart produces and **diastolic pressure** is the blood pressure between heart beats.
- Different factors can cause a person's blood pressure to increase or decrease:
  - It can be increased by stress, high **alcohol** intake, smoking and being overweight.
  - It can be decreased by regular exercise and eating a balanced **diet**.

Remember!
The contraction of heart muscle forces blood through arteries under pressure.

Fitness and health

D–C

- There is a difference between fitness and health:
  - Fitness is the ability to do physical activity.
  - Health is being free from diseases such as those caused by **bacteria** and viruses.
- Your general level of fitness can be measured by your cardiovascular **efficiency**.
- Your fitness can also be measured for different activities:
  - strength, by the amount of weights lifted
  - flexibility, by the amount of joint movement
  - stamina, by the time of sustained exercise
  - agility, by changing direction many times
  - speed, by a sprint race.
- Therefore, you can be very fit for a sprint race but not perform well in a marathon.

Smoking

D–C

- Smoking can increase blood pressure in a number of ways:
  - **Carbon monoxide** in cigarette smoke causes the blood to carry less oxygen. This means that the heart rate increases so that the tissues receive enough oxygen.
  - Nicotine in cigarette smoke directly increases heart rate.

Diet and heart disease

G–E

- The risk of developing heart disease can be increased by high blood pressure, smoking, and eating high levels of salt and saturated fat.
- **Cholesterol** can restrict or block blood flow in arteries by forming **plaques**.
- Analysis of data shows that the incidence of heart disease in the UK is changing.

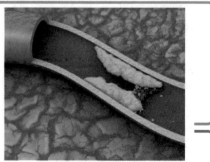

A plaque in an artery

D–C

- Heart disease is caused by a restricted blood flow to the heart muscle. The risk of getting heart disease is increased by:
  - a high level of saturated fat in the diet, which leads to a build-up of cholesterol (a plaque) in arteries
  - high levels of salt, which can increase blood pressure.
- It is essential to be able to interpret data showing links between the amount of saturated fat eaten, the build-up of cholesterol and the incidence of heart disease.

Improve your grade

Healthy and fit?
Joe thinks he is healthy but not fit. Explain the difference. *AO1* [3 marks]

Human health and diet

A balanced diet

- A balanced **diet** should include:
  - protein for growth and repair
  - carbohydrates and fats as high **energy** sources
  - minerals, such as iron to make **haemoglobin**
  - vitamins, such as vitamin C to prevent scurvy
  - fibre to prevent constipation
  - water to prevent **dehydration**.

glucose units

starch is a complex carbohydrate

G–E

amino acids

protein

glycerol   fatty acids

fat

Chemistry of foods

D–C

- It is important for good health to eat a balanced diet containing the correct amounts of the chemicals found in food. Three of these are:
  - carbohydrates, which are made up of simple sugars such as glucose
  - proteins, which are made up of **amino acids**
  - fats, which are made up of fatty acids and glycerol.

- A balanced diet varies according to factors including age, gender, level of activity, religion, being **vegetarian** or **vegan**, or because of medical issues such as food allergies.

Protein intake

- Proteins are only used by the body as an energy source when fats or carbohydrates are not available.

- A high-protein diet is necessary for teenagers, as they are growing fast.

- In many parts of the world, diets are deficient in protein. This is because foods high in protein are in short supply or are too expensive.

G–E

- Proteins are needed for growth and so it is important to eat the correct amount. This is called the estimated average daily requirement (**EAR**) and can be calculated using the formula:

EAR in g = 0.6 × body mass in kg

- Sue has a mass of 72.5 kg. Her EAR is 0.6 × 72.5 = 43.5 g/day.

- Too little protein in the diet causes the condition called **kwashiorkor**. This is more common in developing countries due to overpopulation and lack of money to improve agriculture.

D–C

Overweight or underweight?

- Being overweight (obese) is linked to increased health risks for arthritis, heart disease, diabetes and breast cancer.

G–E

- To work out if a person is overweight or underweight, calculate their **body mass index** (BMI):

$$BMI = \frac{\text{mass in kg}}{(\text{height in m})^2}$$

Tom is 170 cm tall and has a mass of 80 kg. Calculate his BMI.

170 cm = 1.7 m

$$BMI = \frac{80}{1.7^2} = 27.7$$

EXAM TIP
The height must be in metres and the mass in kilograms.

D–C

- A BMI of more than 30 means the person is **obese**, 25–30 is overweight, 20–25 is normal, less than 20 is underweight. With a BMI of 27.7, Tom is overweight.

- Some people may become ill as they choose to eat less than they need. This may be caused by low self-esteem, poor self-image or a desire for what they think is perfection.

Improve your grade

Balanced diet

Charlie is 14 and does athletics. She eats the same diet as her mother. Her mother's diet is a balanced one but Charlie's is not. Explain why. AO1/2 [4 marks]

Staying healthy

The fight against illness

G–E

- Some diseases are infectious (they can be spread to other people). Other diseases, such as **cancer** and inherited disorders, are non-infectious.

- Infectious diseases are caused by **pathogens**. Examples are flu (caused by viruses), cholera (caused by a **bacterium**), athlete's foot (caused by a fungus) and malaria (caused by a protozoan).

- The body defends itself against pathogens: the skin acts as a barrier; blood clots prevent entry of pathogens; mucus in airways traps pathogens; hydrochloric acid kills pathogens in the stomach.

- Pathogens that enter the body are destroyed by **white blood cells**, which are part of the body's immune system. The white blood cells engulf pathogens and **antibodies** destroy them.

- Data is available that links diseases in different countries to the climate as well as to social and economic factors.

D–C

- Pathogens produce the symptoms of an infectious disease by damaging the body's cells or producing poisonous waste products called **toxins**.

- Antibodies lock onto **antigens** on the surface of pathogens such as a bacterium. This kills the pathogen.

- Human white blood cells produce antibodies, resulting in **active immunity**. This can be a slow process but has a long-lasting effect. Vaccinations using antibodies from another human or animal result in passive immunity, which has a quick but short-term effect.

1 this white blood cell recognises bacteria

bacteria

2 antibodies produced

3 antibodies stick bacteria together

4 a different type of white blood cell engulfs bacteria

Destroying pathogens

Malaria

D–C

- Malaria is caused by a protozoan called *Plasmodium*, which feeds on human **red blood cells**.

- *Plasmodium* is carried by mosquitoes, which are **vectors** (i.e. not affected by the disease), and transmitted to humans by mosquito bites.

- *Plasmodium* is a **parasite** and humans are its host. A parasite is an organism that feeds on another living organism, causing it harm.

Cancers

D–C

- Changes in lifestyle and **diet** can reduce the risk of some cancers:
  - Not smoking reduces the risk of lung cancer.
  - Eating fruit and vegetables reduces the risk of bowel cancer.
  - Using sunscreen reduces the risk of skin cancer.

Remember!
Different antibodies are required to deal with different pathogens.

Treatments and trials

G–E

- New medical treatments and new drugs are tested before use to find out if they work and are safe to use.

D–C

- They are tested using animals, human tissue and computer models before human **trials**. Some people object to causing suffering in animals in such tests.

- **Antibiotics** (against bacteria and fungi) and **antiviral drugs** (against viruses) are specific in their action.

- An antibiotic destroys a pathogen; an antiviral drug slows down the pathogen's development.

EXAM TIP
Don't be confused between antibodies and antigens: antibodies lock on to antigens.

Improve your grade

Drug trials

Some scientists believe they have discovered a new drug to treat malaria. Suggest why it may take up to 10 years before the drug is available for general use. *AO2* [4 marks]

The nervous system

Sense organs

- Humans have different types of sense organs to react to different stimuli.

Sense organ (receptor)	Sense	Stimulus
skin	touch	pressure, temperature
tongue	taste	chemicals in food
nose	smell	chemicals in air
eyes	sight	light
ears	hearing and balance	sound

G–E

How do eyes work

- Monocular vision is seeing objects with one eye, **binocular vision** is seeing objects with both eyes.

- Monocular vision has a wider field of vision but causes poor judgement of distance.

- Binocular vision has a narrower field of vision but gives good judgement of distance.

- The main parts of the eye are the cornea, pupil, lens, retina, optic nerve and blind spot.

- The main parts of the eye have special functions.

- Light rays are **refracted** (bent) by the cornea and lens.

- The retina contains light receptors. Some are sensitive to different colours.

- Binocular vision helps to judge distance by comparing the images from each eye; the more different they are, the nearer the object.

G–E

D–C

ciliary muscles control suspensory ligaments

convex lens refracts light rays and focuses light onto the retina

outer cornea refracts light rays

a focused image forms on the retina, which is sensitive to light

pupil allows light rays to enter the eye

coloured iris controls the amount of light entering the eye

optic nerve carries nerve impulses to the brain

suspensory ligaments alter the shape of the lens in focusing

Functions of parts of the eye

Faults in vision

- The main faults in vision are long sight (inability to see near objects clearly), short sight (inability to see distant objects clearly) and red–green colour blindness (inability to distinguish between some colours).

- Red-green colour blindness is caused by a lack of specialised cells in the retina.

- In long sight and short sight, the eyeball or lens is the wrong shape so the image is not focused on the retina.

G–E

D–C

Remember!
In the eye the light rays are refracted, *not* reflected.

Nerve cells

- The nervous system consists of the **central nervous system** (CNS) made up of the brain and spinal cord, and the **peripheral nervous system** (PNS).

- **Reflex** actions, such as the knee jerk, are fast, automatic and protective responses. Voluntary responses, such as picking up a pen, are under conscious control of the brain.

- Nerve impulses (electrical signals) pass along the **axon** of a neurone.

- What happens in a reflex action is shown by a reflex arc. The links in a reflex arc are:

  stimulus→receptor→**sensory neurone**→central nervous system→**motor neurone**→effector→response

- The pathway for a spinal reflex is: receptor→sensory neurone→relay neurone→motor neurone→effector

G–E

D–C

branching dendrites

muscle fibres (effector)

cell body

axon

nucleus

sheath

a motor neurone

Parts of a motor neurone

Improve your grade

Monocular and binocular vision

Predators such as owls hunt and catch their prey. They have good binocular vision. Explain the advantages and disadvantages of binocular vision compared with monocular vision. *AO1* [4 marks]

Drugs and you

Types of drugs

G–E

- Drugs can be beneficial or harmful.

- Some drugs are only available on prescription, as misusing them could cause harm.

- Being addicted to a drug means it is hard to give it up. After taking a drug for a while, the person becomes tolerant of it, so they need larger doses for the same effect. When trying to give up a drug, a person may experience **withdrawal symptoms**, such as being irritable. Having treatment to help recover from drug addition is called rehabilitation.

- Each drug category has different effects:
  - **Depressants** slow down the brain's activity.
  - **Painkillers** block nerve impulses.
  - **Stimulants** increase the brain's activity.
  - **Performance enhancers** increase muscle development.
  - **Hallucinogens** distort what is seen and heard.

D–C

- Drugs have a legal classification. Class A drugs are the most dangerous and have the heaviest penalties. Class C drugs are the least dangerous, with the lightest penalties.

- There are different types of drugs in the different categories, for example:
  - depressants (**alcohol**, solvents, temazepam)
  - painkillers (aspirin, paracetamol)
  - stimulants (nicotine, MDMA ('ecstasy'), caffeine)
  - performance enhancers (anabolic steroids)
  - hallucinogens (LSD).

> **Remember!**
> Alcohol is a depressant.
> Caffeine is a stimulant.

Effects of smoking

G–E

- Tobacco smoking can cause emphysema, bronchitis, heart disease and cancer of the mouth, throat, oesophagus and lung.

- Tobacco smoke contains:
  - **carbon monoxide**, which causes a lack of oxygen and heart disease
  - nicotine, which is addictive
  - tars, which irritate and cause cancer
  - particulates, which accumulate in lung tissue.

D–C

- Cigarette smoke contains many chemicals that stop cilia moving.

- Cilia (tiny hairs) are found in the epithelial lining of the trachea, bronchi and bronchioles.

- A 'smoker's cough' is a result of:
  - dust and particulates in cigarette smoke collecting and irritating the epithelial lining
  - mucus not being moved by the cilia.

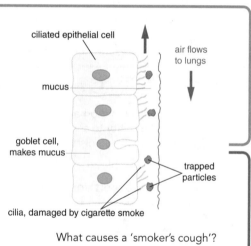

ciliated epithelial cell

air flows to lungs

mucus

goblet cell, makes mucus

trapped particles

cilia, damaged by cigarette smoke

What causes a 'smoker's cough'?

Effects of alcohol

G–E

- The short-term effects of alcohol on the body include impaired judgement, balance and muscular control, blurred vision, slurred speech, drowsiness, increased blood flow to skin.

- Because of these effects, drinking alcohol before driving a car or piloting a plane is extremely dangerous. There is therefore a legal limit for the level of alcohol in the breath/blood for drivers and pilots.

- Drinking excess alcohol has long-term effects, such as liver and brain damage.

D–C

- The alcohol content of alcoholic drinks is measured in **units of alcohol**.

- Drinking alcohol increases **reaction times** and increases the risk of accidents.

Improve your grade

Smoking

Asif smokes cigarettes. Suggest why he finds it difficult to give up smoking and list the extra risks of being a smoker. *AO1* [4 marks]

B1 Understanding organisms

Staying in balance

Homeostasis

- The body maintains steady levels of **temperature**, water and **carbon dioxide**. This is essential to life.

- Keeping a constant internal environment is called homeostasis.

- Homeostasis involves balancing bodily inputs and outputs.

- Automatic control systems keep the levels of temperature, water and carbon dioxide steady. This makes sure all cells can work at their optimum level.

G–E

D–C

Temperature control

- The core body temperature is normally kept at approximately 37 °C.

- Body temperature can be measured accurately in the mouth, anus and ear.

- Clinical thermometers, sensitive strips, digital recording probes and thermal imaging can be used to measure body temperature.

- Heat can be gained/retained by **respiration** (releasing **energy** from food), shivering (heat from muscles working), exercise, less sweating, less blood flow near skin surface and wearing more clothing.

- Heat may be lost by sweating (energy used in evaporation) and by more blood flow near the skin surface.

G–E

- The body temperature of 37 °C is linked to the **optimum temperature** for many **enzymes**.

- A high temperature can cause:
  - **heat stroke** (skin becomes cold and clammy and pulse is rapid and weak)
  - **dehydration** (loss of too much water).

- Both heat stroke and dehydration can be fatal if not treated.

- To avoid overheating, sweating increases heat transfer from the body to the environment.

- The **evaporation** of sweat requires body heat to change the liquid sweat into water vapour.

- A very low temperature can cause **hypothermia** (slow pulse rate, violent shivering), which can be fatal if not treated.

Remember!
Heat stroke, dehydration and hypothermia can be fatal.

D–C

Control of blood sugar levels

- The pancreas (which sits just under and partly behind the stomach) produces the **hormone** called **insulin**.

- Type 1 diabetes is caused by the failure of the pancreas to produce insulin.

- Insulin travels around the body in the blood.

G–E

- Insulin controls **blood sugar levels**.

- Hormone action is slower than nervous reactions as the hormones travel in the blood.

- Type 1 diabetes must be treated by insulin. Type 2 diabetes, which is caused either by the body producing too little insulin or the body not reacting to it, can be controlled by **diet**.

D–C

Improve your grade

Diabetes

Describe and explain the link between the pancreas and Type 1 diabetes.
AO1/2 [4 marks]

Controlling plant growth

Plant responses

- Plants as well as animals respond to changes in their environment.

- The growth of shoots and roots, flowering and fruit ripening is controlled by chemicals called plant hormones.

- A shoot's growth towards light increases the amount of glucose made by the plant during **photosynthesis**. This increases the plant's chance of survival.

- An experiment such as growing cress seedlings shows how plant shoots grow towards the light.

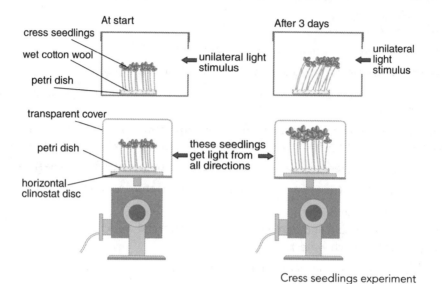

Cress seedlings experiment

- A root's growth towards gravity increases the stability of the plant and its chances of finding water.

- **Phototropism** is a plant's growth response to light. **Geotropism** is a plant's growth response to **gravity**.

- Parts of a plant respond in different ways:
  - Shoots are positively phototropic (they grow towards light) and negatively geotropic (they grow away from the pull of gravity).
  - Roots are negatively phototropic (they grow away from light) and positively geotropic (they grow with the pull of gravity).

> **Remember!**
> A positive reaction means that the root or shoot grows towards the stimulus.

Plant hormones (auxins)

- **Auxins** are a group of plant hormones. They move through the plant in solution.

- Auxins are involved in phototropism and geotropism.

Commercial uses of plant hormones

- Plant hormones can be used in agriculture to speed up or slow down plant growth.

- Plant hormones are used:
  - as selective weedkillers
  - rooting powder when taking cuttings
  - to delay or accelerate fruit ripening
  - to control dormancy in seeds.

Improve your grade

Phototropism

Look at the diagram above showing an experiment on cress seedlings.
What is the experiment designed to find out? Explain the plants' reactions.
AO2 [4 marks]

Variation and inheritance

Inherited characteristics

- Some human characteristics, such as intelligence, body mass and height, are the result of both environmental and inherited factors.

- Some human characteristics, such as facial features, can be inherited. They can be **dominant** or **recessive**.

- **Alleles** are different versions of the same **gene**.

G–E

D–C

Chromosomes

- **Chromosomes** are found in the nucleus of a cell.

- Chromosomes carry information in the form of genes, which control inherited characteristics.

- Most body cells contain matching pairs of chromosomes.

- **Gametes** (eggs and sperm) have half the number of chromosomes found in body cells.

Contents of a cell

cell membrane

nucleus

cytoplasm

chromosomes carry genetic information in genes

G–E

- Most body cells have the same number of chromosomes. The number depends on the **species** of organism. Human cells have 23 pairs.

- **Sex chromosomes** determine sex in **mammals**. Females have identical sex chromosomes called XX, males have different sex chromosomes called XY.

- A sperm will carry either an X or a Y chromosome. All eggs will carry an X chromosome.

- There is a **random** chance of which sperm **fertilises** an egg. There is therefore an equal chance of the offspring being male or female.

X X
female

X Y
male

eggs

sperm

X X
female

X Y
male

X X
female

X Y
male

Inheritance of sex

D–C

Genetic variation

- Genetic **variation** is caused by:
  - **mutations**, which are random changes in genes or chromosomes
  - rearrangement of genes during the formation of gametes
  - **fertilisation**, which results in a zygote with alleles from the father and mother.

D–C

Inherited disorders

- Some disorders, such as red–green colour blindness, sickle cell anaemia and cystic fibrosis, are inherited.

- Inherited disorders are caused by faulty genes.

- Many personal and ethical issues are raised:
  - in deciding to have a genetic test (a positive result could alter lifestyle, career, insurance)
  - by knowing the risks of passing on an inherited disorder (whether to marry/have a family).

G–E

D–C

Improve your grade

Breeding tomato plants

Look at the diagram of a breeding experiment using tomato plants.

(a) Where is the genetic material about stem colour held in the tomato plants?

(b) Explain the results of the F1 generation.

AO1/2 [5 marks]

Parental generation

purple-stemmed   green-stemmed

F1 generation

purple-stemmed

B1 Summary

A balanced diet varies according to age, gender, activity, religion and personal choice.

The EAR can be used to calculate protein requirements

EAR in g = body mass in kg × 0.6

A balanced diet will vary according to age, gender, activity, religion and personal choice.

$$BMI = \frac{mass\ in\ kg}{(height\ in\ m)^2}$$

Being fit is the ability to do exercise, being healthy is being free from disease.

Harmful drugs are classified as Class A, B and C. Class A is the most harmful. Depressant and stimulant drugs affect the nervous system.

Diet, drugs and disease

Immunisation protects against certain diseases by using harmless pathogens.

Blood pressure has two readings: diastolic and systolic pressure in mmHg.

Smoking, a high alcohol intake and a diet rich in saturated fats and salt all increase blood pressure.

High blood pressure can damage the brain and kidneys.

The mosquito is a vector that carries malaria. *Plasmodium* is the pathogen that causes malaria. It is a parasite and humans are its host.

Homeostasis is maintaining a constant internal environment.

Automatic systems in the body keep water, temperature and carbon dioxide levels constant.

The hormone insulin controls blood sugar levels. Type 1 diabetes is caused by a failure of the pancreas to make insulin.

Homeostasis and plant hormones

Auxins are a group of plant hormones.

Auxins are involved in phototropism (response to light) and geotropism (response to gravity).

Plant hormones have many commercial uses (selective weedkiller, rooting powder, control of fruit ripening).

Light rays are refracted as they pass through the cornea and lens.

The eye accommodates by altering the shape of the lens.

Long and short sight is caused by the eyeball or lens being the wrong shape.

The nervous system

Monocular vision has a wider field of view but poorer distance judgement than binocular vision.

A nerve impulse is an electrical signal.

Reflex actions are fast, automatic and protective.

A spinal reflex involves a receptor, sensory, relay and motor neurones and an effector.

Alleles are different versions of the same gene. Inherited disorders are caused by faulty alleles.

Most faulty alleles are recessive.

Sex is determined by sex chromosomes, XX in female, XY in male.

Variation and inheritance

Intelligence, body mass and height are controlled by genetic and environmental factors.

Chromosomes are in the nucleus of a cell. They carry information in the form of genes.

Human body cells have 23 pairs of chromosomes.

Classification

Grouping organisms

- For thousands of years scientists have been putting organisms into groups or classifying them according to how similar or how different they are.

- The first step in a classification system is to put an organism into a kingdom. At present, five kingdoms are used:
  - Plants all have a cellulose cell wall and use light **energy** to produce food.
  - Animals are all multicellular and feed on other organisms.
  - Fungi all have a cell wall made of chitin and produce spores.
  - Protoctista are mostly singled-celled organisms with a nucleus.
  - Prokaryotes are single celled and do not have a nucleus.

- For example, if we want to classify a spider:
  - First, it would be put in the animal kingdom, as it is multicellular and feeds on other organisms.
  - Then it would then be classified as an arthropod, as it has jointed legs and an external skeleton.
  - The arthropods are then divided into four smaller groups called classes. These are insects, arachnids, crustaceans and myriapods.
  - A spider is an arachnid, as it has two parts to the body and eight legs.

G–E

- All organisms are classified into a number of different groups, starting with their kingdom and finishing with their **species**.

- The groups are: kingdom, phylum, class, order, family, genus and species.

- As you move down towards 'species', there are fewer organisms within each group and they share more similarities.

Remember!
Try remembering the order of the groups by using the first letter of each one.

D–C

Species

- Humans are all classified in one species.
  - This means that all humans have more similarities with each other than they share with other organisms like chimpanzees.
  - There is, however, quite a lot of **variation** between individual humans, even though we are all in the same species.

G–E

- A species is a group of organisms that can interbreed to produce fertile offspring.

- All organisms are named by the **binomial system**. The system works like this:
  - There are two parts to the name, the first is the genus and the second the species.
  - The genus part starts with a capital letter; the species part starts with a lower-case letter.
  - The same names are used in all countries which avoids confusion.

D–C

Problems with classifying

- Living things are at different stages of **evolution**, and new ones are being discovered all the time. This makes it difficult to place organisms into distinct groups. An example of this is *Archaeopteryx*. This creature had characteristics that would put it into two different groups:
  - It had feathers, like a bird.
  - It also had teeth and a long, bony tail, like a **reptile**.

D–C

Classification and evolution

- Organisms that are grouped together are usually closely related. This means that:
  - They often live in similar **habitats** as they have similar requirements.

G–E

  - They may share a recent common ancestor.
  - They may have different features if they live in different habitats.

D–C

Improve your grade

Classifying newly discovered organisms
A new group of animals is discovered. What steps would scientists take to work out if it is a new species and to give it a name? *AO2* [4 marks]

Energy flow

Pyramids of biomass

- A food web shows the feeding relationships between organisms in a **habitat**.

- Each stage or feeding level in a food chain or food web is called a **trophic level**.

- **Producers** are at the start of a food web because they can make their own food.
  - Most are green plants or algae that make food by **photosynthesis**.
  - A very small number are **bacteria**, which make food using **energy** from chemical reactions.

- Very few organisms eat only one type of food. Most will eat several types and the food might be from different trophic levels. For example, in this food web the hedgehog eats both grass and beetles. When it eats grass it is a primary consumer and when it eats beetles it is a secondary consumer.

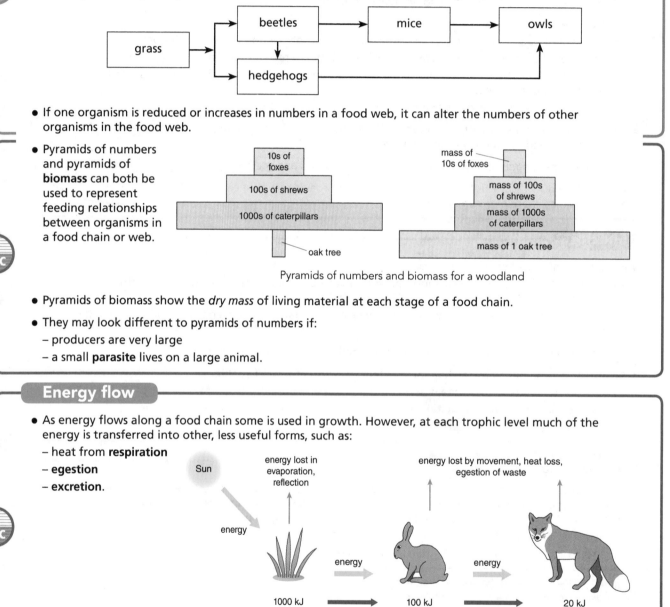

- If one organism is reduced or increases in numbers in a food web, it can alter the numbers of other organisms in the food web.

- Pyramids of numbers and pyramids of **biomass** can both be used to represent feeding relationships between organisms in a food chain or web.

Pyramids of numbers and biomass for a woodland

- Pyramids of biomass show the *dry mass* of living material at each stage of a food chain.

- They may look different to pyramids of numbers if:
  - producers are very large
  - a small **parasite** lives on a large animal.

Energy flow

- As energy flows along a food chain some is used in growth. However, at each trophic level much of the energy is transferred into other, less useful forms, such as:
  - heat from **respiration**
  - **egestion**
  - **excretion**.

Energy flow in a food chain

- The material that is lost at each stage of the food chain is not wasted. Most of the waste is used by **decomposers** that can then start another food chain.

Improve your grade

Pyramids of biomass
It is sometimes hard to work out which trophic level to place an organism in when drawing a pyramid of numbers. Even if it is possible, it does not always form a pyramid. Explain these observations. *AO1* [4 marks]

Recycling

The carbon cycle

- When plants and animals die and **decay**, the elements in their bodies can be **recycled**. Recycling is important to avoid supplies of some elements running out.

- **Carbon** is one of a number of elements that are found in living organisms:
  - Like other elements, carbon therefore needs to be recycled so it can become available again to other living organisms.
  - This recycling happens when animals and plants decay.
  - This process of decay is carried out by **bacteria** and fungi acting as **decomposers**.

- Carbon can be taken up by plants as **carbon dioxide** from the air.

- This carbon dioxide is used by plants for **photosynthesis**.

- Feeding passes carbon compounds along a food chain or web.

- Carbon dioxide is released into the air by:
  - plants and animals **respiring**
  - soil bacteria and fungi acting as decomposers
  - the burning of **fossil fuels (combustion)**.

G–E

D–C

Remember!

Carbon dioxide can be locked up in limestone for a long time. The oceans are often called a carbon sink.

The nitrogen cycle

- Nitrogen makes up about 78% of the air but it is very unreactive, so cannot be used directly by plants.

- Plants therefore have to get their nitrogen as nitrates from the soil.

- Plants use nitrogen to make protein for growth.

- Feeding passes nitrogen compounds along a food chain or web.

- The nitrogen compounds in dead plants and animals are broken down by decomposers and returned to the soil.

G–E

D–C

EXAM TIP

Diagrams of the carbon or nitrogen cycle may not look exactly like this one. Just look for where the arrows go to and from.

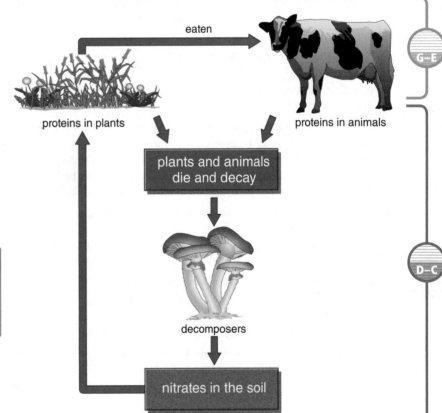

eaten

proteins in plants

proteins in animals

plants and animals die and decay

decomposers

nitrates in the soil

The nitrogen cycle

Keeping decomposers working

- For decomposers to break down dead material in soil, they need oxygen and a suitable **pH**:
  - Decay will therefore be slower in waterlogged soils as there will be less oxygen.
  - Acidic conditions will also slow down decay.

D–C

Improve your grade

The nitrogen cycle

Explain how nitrogen in a protein molecule in a dead leaf can become available again to a plant. *AO1* [4 marks]

Interdependence

Competition

G–E

- Organisms that are adapted to their **habitats** are better able to compete for resources.
- Organisms are not spread evenly in all habitats:
  - There might be habitats where some organisms do not live.
  - In some habitats the number of individuals might be much higher than in other habitats.
- Many of the differences in numbers are due to how successful organisms are in competing with each other for food, shelter, water, light or minerals.

D–C

- Similar animals living in the same habitat compete with each other for resources (e.g. food).
- If they are members of the same **species** they will also compete with each other for mates so they can breed.

Predator–prey relationships

G–E

- **Predators** eat **prey** and so can reduce the number of prey organisms that survive.

D–C

- Both predator and prey show cyclical changes (ups and downs) in their numbers. This is because:
  - When there are lots of prey, more predators survive and so their numbers increase.
  - The increased number of predators eat more prey, so prey numbers drop.
  - More predators starve and so their numbers drop.

Parasitism and mutualism

G–E

- As well as competing with each other or eating each other, organisms can also be dependant on organisms from a different species. This might be in a number of ways.
- Sometimes both organisms benefit as a result of their relationship.
  - Buffalo and oxpecker birds live together. The birds eat insects that feed on the buffalo so both the buffalo and the bird gain.
  - On some coral reefs, 'cleaner' fish are regularly visited by larger fish. The larger fish benefit by having their **parasites** removed by the cleaner fish and the cleaner fish gain food.

D–C

- When both organisms benefit as a result of their relationship, this is called **mutualism**:
  - Insects visit flowers and so transfer pollen, allowing **pollination** to happen. They are 'rewarded' by sugary nectar from the flower.
- Parasites feed on or in another living organism called the host:
  - The host suffers as a result of the relationship.
  - Fleas are parasites living on a host (which may be human).
  - Tapeworms are also parasites feeding in the digestive systems of various animals.

Remember!
A successfully adapted parasite does not kill its host quickly, as it would then need to find another one.

Improve your grade

Fish and shrimp

On coral reefs there are areas where shrimps live. Fish regularly visit these areas and allow the shrimps to move over their bodies, feeding.

What type of relationship is this and what is the benefit to each organism?
AO2 [2 marks]

Adaptations

Adaptions of predators and prey

- Some animals called **predators** are adapted to hunt other animals for food. The animals that are hunted are called **prey** and are adapted to help them to escape.
- Predators may be adapted by:
  - having eyes on the front of their head, which gives **binocular vision** to judge size and distance
  - hunting in groups, so that they can surround prey or bring down larger prey
  - producing their offspring at a time when there is plenty of food available.
- Prey animals are adapted by:
  - having eyes on the side of head to give a view all around
  - having a body that is coloured to give a warning that it is poisonous or has a sting
  - looking like a poisonous animal even if it is harmless (mimicry)
  - living in groups such as herds or shoals which reduce the chance of being caught
  - all breeding at the same time so that at least some of the offspring will survive (synchronous breeding).

G–E

Adapting to the cold

- Some animals are adapted to living in very cold conditions. They keep warm by reducing heat loss. Some have anatomical **adaptations** to help reduce heat loss.
- So:
  - They have excellent **insulation** to cut down heat loss. The arctic fox has thick fur that traps plenty of air for insulation. Seals have thin fur but a thick layer of fat under the skin.
  - These animals are usually quite large, with small ears. This helps to decrease heat loss by decreasing the surface area to volume ratio.
- Animals may try to avoid the cold by changing their behaviour. Some migrate long distances to warmer areas. Others slow down all their body processes and hibernate.

D–C

Adapting to hot, dry conditions

- Organisms such as camels and cacti live in deserts, in very hot, dry conditions.
- To increase heat loss, animals adapt in a variety of ways:
  - Some are anatomical adaptations, for example camels increase the loss of heat by having very little hair on the underside of their bodies. Animals that live in hot areas are usually smaller and have larger ears than similar animals that live in cold areas. These factors give them a larger surface area to volume ratio, so that they can lose more heat.
  - Other adaptations to lose more heat are behavioural, such as panting or licking their fur.
- To reduce heat gain, animals may change their behaviour, for example they seek shade during the hotter hours around the middle of the day.
- To cope with dry conditions, organisms have behavioural, anatomical and physiological adaptations. For example:
  - Camels can survive with little water because they can produce very concentrated urine.
  - Cacti reduce water loss because their leaves have been reduced to spines. They also have deep roots and can store water in the stem.

Spines on a cactus

D–C

Improve your grade

Living in hot, dry conditions

A camel has wide feet, skin with few hairs underneath the body and the ability to produce concentrated urine. Suggest how these features are an advantage to living in hot, dry areas. *AO2* [3 marks]

EXAM TIP

You may have to write about different organisms. Just apply the same principles.

Natural selection

Charles Darwin and natural selection

G–E

- **Evolution** is the gradual change in a group of organisms over a long period of time.
- When the environment that organisms live in changes, groups of organisms may respond differently:
  - Some groups of organisms may evolve and become better adapted and survive.
  - Others may become **extinct**.

EXAM TIP
Do not talk about individual organisms evolving. It is groups of organisms that evolve or become extinct.

D–C

- Over 150 years ago Charles Darwin wrote his theory of **natural selection** to explain how evolution might happen. It says that if animals and plants are better adapted to their environment, they and the following generations are more likely to survive.
- He did not know exactly how **adaptations** were passed on. We now know that when organisms reproduce, their **genes** are passed on to the next generation.
- The modern version of natural selection can be summarised like this:
  - Within any **species** there is **variation**.
  - Organisms produce far more young than will survive, so there is competition for limited resources such as food.
  - Only those best adapted will survive, which is called *survival of the fittest*.
  - Those that survive pass on successful adaptations to the next generation in their genes.

Modern examples of natural selection

D–C

- Natural selection is difficult to study because it usually takes thousands of years to see the effects. Some examples have been studied over shorter time spans:
  - More and more **bacteria** are developing resistance to **antibiotics**.
  - Peppered moths are dark or pale in colour. Dark moths are better camouflaged in polluted areas, so more of them survive.

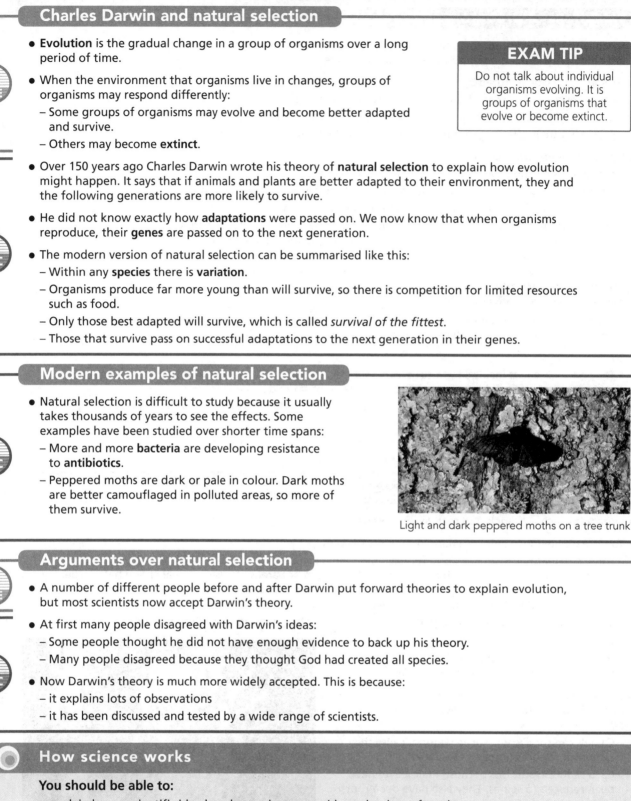

Light and dark peppered moths on a tree trunk

Arguments over natural selection

G–E

- A number of different people before and after Darwin put forward theories to explain evolution, but most scientists now accept Darwin's theory.

D–C

- At first many people disagreed with Darwin's ideas:
  - Some people thought he did not have enough evidence to back up his theory.
  - Many people disagreed because they thought God had created all species.
- Now Darwin's theory is much more widely accepted. This is because:
  - it explains lots of observations
  - it has been discussed and tested by a wide range of scientists.

How science works

You should be able to:

- explain how a scientific idea has changed as new evidence has been found
- describe how a famous scientist made a series of observations in order to develop new scientific explanations
- recognise that confidence increases in scientific explanations if observations match predictions.

Improve your grade

Explanations for evolution

Human ancestors had more hair than modern humans. One idea why modern humans have less hair is that it might make them less likely to have parasites. Use Darwin's ideas to explain why modern humans have less hair than their ancestors. *AO2* [4 marks]

Population and pollution

Pollution

- There are many different types of **pollution**. Three that have caused much concern are:
  - **carbon dioxide**, from increased burning of **fossil fuels**, which may increase the greenhouse effect and **global warming**
  - **CFCs**, from aerosols, which destroy the **ozone layer**
  - sulfur dioxide, from burning fossil fuels, which causes **acid rain**.

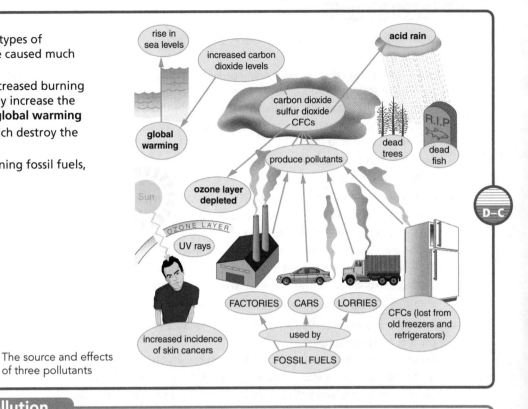

The source and effects of three pollutants

Population and pollution

- The **population** of the planet is increasing and so there is a greater demand for certain resources.

- Some of these resources are in limited supply (**finite**). These include:
  - minerals
  - fossil fuels.

- The increasing human population has also produced more waste. This includes:
  - household rubbish or waste
  - sewage from flushed toilets
  - sulfur dioxide and carbon dioxide from burning fossil fuels.

- The human population of the world is growing at an ever-increasing rate. This is called **exponential growth**.

- This growth in population is happening because the birth rate is exceeding the death rate.

Measuring pollution

- Pollution can change the number or type of organisms that live in a particular area.

- Pollution in water or air can be measured using direct methods or by indicator organisms.

- Direct methods include oxygen probes attached to computers that can measure the exact levels of oxygen in a pond. Special chemicals can be used to indicate levels of nitrate pollution from **fertilisers**.

- The presence or absence of an **indicator species** is used to estimate levels of pollution. For example:
  - The mayfly larva is an insect that can only live in clean water.
  - The water louse, bloodworm and mussels can live in polluted water.
  - Lichen grows on trees and rocks but only when the air is clean. It is unusual to find lichen growing in cities, because it is killed by the pollution from motor engines.

Remember!
If many different species are present in a habitat, this is usually a sign of low levels of pollution.

Improve your grade

Population and pollution
Explain the reasons for the increase in carbon dioxide levels in the atmosphere and explain why people are concerned about this. *AO1* [4 marks]

Sustainability

Conservation

- **Conservation** involves trying to preserve the variety of plants and animals and the **habitats** that they live in.
- Conservation is needed because many organisms have become **extinct** or are **endangered**. This may be for any of the following reasons:
  - The climate has changed, making them less suited to the conditions.
  - Their habitat has been destroyed, so that people can use the land.
  - People have hunted them.
  - Pollution has killed them.
  - Other organisms have been introduced and competed with them.

- People have tried to help endangered **species** survive in a number of different ways by:
  - passing laws that stop their habitats being destroyed
  - passing laws to make hunting illegal
  - telling people about the organisms so they are more aware of the problems
  - breeding the animals in captivity, such as in zoos, so they can be released into the wild
  - setting up seed banks that store the seeds of rare plants
  - creating artificial ecosystems where the organisms can live in safety.

- People think that conservation is important because it can:
  - protect our food supply
  - prevent any damage to food chains, which can be hard to predict
  - protect plants and animals that might be useful for medical purposes
  - protect organisms and habitats that people enjoy to visit and study.

- Species are at risk of extinction if the number of individuals or habitats falls below critical levels.

Whale conservation

- Many different whale species are endangered or close to extinction.

- Whales have been hunted for hundreds of years for their body parts, which are used in many products. Live whales are also important to the tourist trade.

- Some whales are kept in captivity for research or **captive breeding** programmes or just for our entertainment. However, many people object when whales lose their freedom.

skin: used in belts, shoes, handbags and luggage

sinews: used in tennis rackets

spermacetti: used in high-grade machine oil

oil: sperm whale oil taken from bone and skin used in high-grade alcohol, shoe cream, lipstick, ointment, crayons, candles, fertiliser, soap and animal feeds

whalemeat: used in pet food and human food

teeth: used in buttons, piano keys and jewellery

liver: used in oil

ambergris: from intestine, used in perfumes

bone: used in fertiliser and animal feeds

Many uses of whale parts

Sustainable development

- A sustainable resource is something that can be taken out of the environment without it running out. Both woodland and fish supplies can be sustainable if they are treated in the right way.

- **Sustainable development** means taking enough resources from the environment for current needs, while leaving enough for the future and preventing permanent damage. For example:
  - Fishing quotas are set, so that there are enough fish left to breed.
  - Woods are replanted to keep up the supply of trees.

Improve your grade

Saving endangered species

The Hawaiian goose is only found on the islands of Hawaii. In the mid-1900s only about 30 were left alive. The space on the islands is restricted and a number of animals have been introduced to the islands. Write about the problems facing scientists trying to save the goose from extinction. *AO2* [4 marks]

B2 Summary

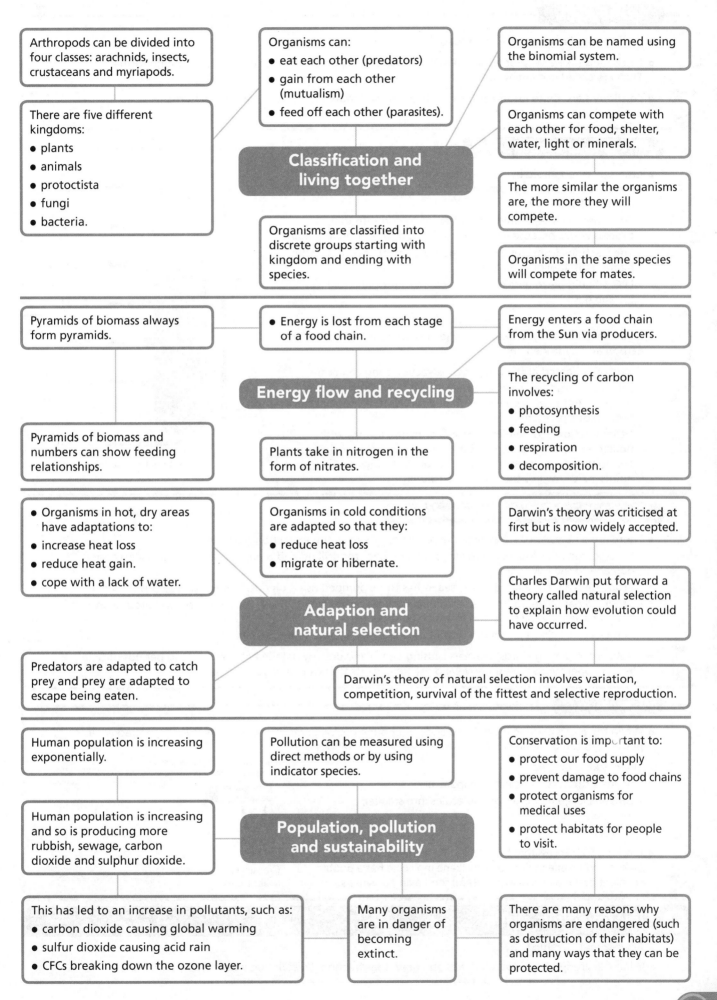

Arthropods can be divided into four classes: arachnids, insects, crustaceans and myriapods.

There are five different kingdoms:
- plants
- animals
- protoctista
- fungi
- bacteria.

Organisms can:
- eat each other (predators)
- gain from each other (mutualism)
- feed off each other (parasites).

Organisms can be named using the binomial system.

Organisms can compete with each other for food, shelter, water, light or minerals.

The more similar the organisms are, the more they will compete.

Classification and living together

Organisms are classified into discrete groups starting with kingdom and ending with species.

Organisms in the same species will compete for mates.

Pyramids of biomass always form pyramids.

- Energy is lost from each stage of a food chain.

Energy enters a food chain from the Sun via producers.

Energy flow and recycling

The recycling of carbon involves:
- photosynthesis
- feeding
- respiration
- decomposition.

Pyramids of biomass and numbers can show feeding relationships.

Plants take in nitrogen in the form of nitrates.

- Organisms in hot, dry areas have adaptations to:
- increase heat loss
- reduce heat gain.
- cope with a lack of water.

Organisms in cold conditions are adapted so that they:
- reduce heat loss
- migrate or hibernate.

Darwin's theory was criticised at first but is now widely accepted.

Charles Darwin put forward a theory called natural selection to explain how evolution could have occurred.

Adaption and natural selection

Predators are adapted to catch prey and prey are adapted to escape being eaten.

Darwin's theory of natural selection involves variation, competition, survival of the fittest and selective reproduction.

Human population is increasing exponentially.

Pollution can be measured using direct methods or by using indicator species.

Conservation is important to:
- protect our food supply
- prevent damage to food chains
- protect organisms for medical uses
- protect habitats for people to visit.

Human population is increasing and so is producing more rubbish, sewage, carbon dioxide and sulphur dioxide.

Population, pollution and sustainability

This has led to an increase in pollutants, such as:
- carbon dioxide causing global warming
- sulfur dioxide causing acid rain
- CFCs breaking down the ozone layer.

Many organisms are in danger of becoming extinct.

There are many reasons why organisms are endangered (such as destruction of their habitats) and many ways that they can be protected.

Making crude oil useful

Fossil fuels

- Coal, gas and **crude oil** are **fossil fuels**.
- Fossil fuels take a very long time to make and are being used up faster than they are being formed. They are called **non-renewable** fuels.
- Fossil fuels are **finite resources** because they are no longer being made, or are being made extremely slowly.

Fractional distillation

- Crude oil is separated by heating it up and then, when parts of it have boiled, cooling it down and collecting the liquid. This is called fractional **distillation**.
  - The crude oil is separated into different fractions (parts). The process works because each fraction has a different **boiling point**.
  - The fractions from crude oil include LPG, **petrol**, diesel, paraffin, heating oil, fuel oils and **bitumen**.
  - LPG is liquid petroleum gas, which contains propane and butane gases.
- Crude oil is a mixture of many types of oil, which are all 'hydrocarbons'. A hydrocarbon is made up of molecules containing **carbon** and hydrogen only.
- Crude oil is heated at the bottom of a fractionating column.
  - Oil that doesn't boil sinks as a thick liquid to the bottom. This is bitumen. Bitumen has a very high boiling point. It 'exits' at the bottom of the column.
  - Other fractions, containing mixtures of hydrocarbons with similar boiling points, boil and their gases rise up the column. The column is cooler at the top. Fractions with lower boiling points 'exit' towards the top of the column.

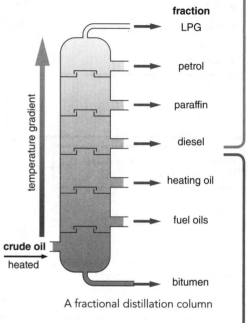

A fractional distillation column

Problems in extracting crude oil

- There are a number of environmental problems linked to the extraction of crude oil.
  - When crude oil is found, a large area is taken over for drilling and pumping oil to the surface, causing damage to the environment.
  - Crude oil is also found under the sea. It has to be pumped using oil rigs. This is a dangerous activity.
  - Crude oil has to be transported through pipelines or by tanker. If sea-going tankers run aground and are damaged, the oil spills and forms an oil slick.
  - This can cause enormous harm to wildlife and beaches have to be cleaned up.
- Oil slicks can damage birds' feathers causing death and destroy unique habitats for long periods of time. Clean-up operations are expensive and use detergents that can damage wildlife.

Cracking

- When too much paraffin is produced and not enough petrol, paraffin can be broken down or '**cracked**' into petrol. This can be done on an industrial scale or in the laboratory.
- The cracking process:
  - needs a **catalyst** and a high **temperature**
  - converts large hydrocarbon molecules into smaller, more useful ones
  - makes more petrol.
- Cracking is a process that turns large **alkane** molecules into smaller alkane and **alkene** molecules. An alkene molecule has a **double covalent bond**, which makes it useful for making polymers.

liquid paraffin (large molecules) on mineral fibre

hydrocarbon gas (small molecules)

aluminium oxide

very strong heat

water

Cracking in a laboratory

Improve your grade

Fractional distillation

What can crude oil be separated into and how does fractional distillation work? *AO1* [4 marks]

Using carbon fuels

Choosing fuels

- A fuel is chosen because of its key features:
  - **energy** value
  - **pollution** caused
  - availability
  - ease of use
  - storage
  - cost.
  - toxicity

Fuel	Able to flow?	How it burns
petrol	yes	instantly
coal	no	glows for a time

G–E

- When this information is known, fuels can be chosen for particular uses.

- The data in the table above helps explain why the Victorians used coal in their fireplaces.

- When a number of features are known, it is easier to choose the best fuel for a purpose. For example, the table below shows that **petrol** flows easily in engines so it is a better fuel for cars.

Key features	Coal	Petrol
energy value	high	high
availability	good	good
storage	bulky and dirty	volatile
cost	high	high
toxicity	produces acid fumes	produces less acid fumes
pollution caused	acid rain, carbon dioxide, soot	carbon dioxide, nitrous oxides
ease of use	easier to store for power stations	flows easily around engines

D–C

Combustion

- When a fuel burns it releases useful heat energy.
  - Oxygen is needed for fuels to burn. **Complete combustion** needs a plentiful supply of oxygen (air).
  - Complete combustion of a hydrocarbon fuel makes **carbon dioxide** and water only.
  - The word equation for fuel burning in air is:

  fuel + oxygen → carbon dioxide + water

- If a fuel burns in a shortage of oxygen, **combustion** is incomplete.

- When a fuel burns in a shortage of oxygen it produces water and unwanted gases.
  - These gases contain soot and are **toxic**.
  - One of the toxic fumes is a gas called **carbon monoxide**.
  - Carbon monoxide is a poisonous gas and is very dangerous if it is breathed in.
  - There are two word equations for incomplete combustion:

  fuel + oxygen → carbon monoxide + water

  fuel + oxygen → carbon + water

- A Bunsen burner flame produces energy from burning gas.
  - More energy is transferred by a blue Bunsen flame than a yellow Bunsen flame.
  - When combustion is incomplete, a yellow flame is seen. This is because a yellow flame produces a lot of soot. Carbon monoxide, soot and water vapour are produced.

Remember!
Complete combustion gives carbon dioxide and water. Incomplete combustion gives carbon monoxide and water or carbon and water.

G–E

- The complete combustion of a hydrocarbon fuel can be shown using an experiment in the laboratory.

- The advantages of complete combustion over incomplete combustion are that:
  - More energy is released during complete combustion than during incomplete combustion.
  - Toxic gas (carbon monoxide) and soot (carbon) are made during incomplete combustion.

to filter pump to draw air through

beaker of cold water

candle

cobalt chloride paper to test for water vapour

conical flask with limewater to test for carbon dioxide

Fuels burn in oxygen to make carbon dioxide and water

D–C

Improve your grade

Choosing a fuel
Use the table at the top of the page to decide which fuel should be used in a car engine. Justify your answer from the evidence. *AO3* [2 marks]

Clean air

What is in clean air?

- Air is a mixture of different gases. Clean air is air that contains no **pollutants** caused by human activity. The main gases in clean air are:
  - nitrogen
  - oxygen
  - **carbon dioxide**
  - water vapour
  - noble gases
    (argon, neon, krypton, xenon).

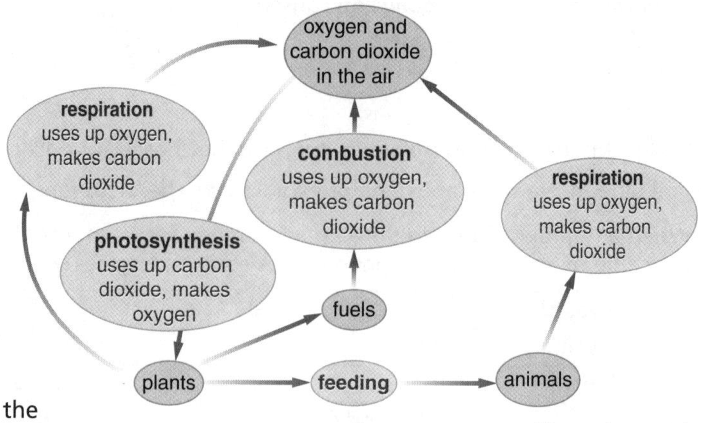

- The amount of water vapour in the air changes considerably. The amounts of oxygen, nitrogen and carbon dioxide in the air remain almost constant.

- The levels of gases in the air depend on:
  - **combustion** of **fossil fuels**, which increases the level of carbon dioxide and decreases the level of oxygen
  - **respiration** by plants and animals, which increases the level of carbon dioxide and decreases the level of oxygen.
  - **photosynthesis** by plants, which decreases the level of carbon dioxide and increases the level of oxygen.

The carbon cycle

- Clean air is made up of 78% nitrogen, 21% oxygen and of the remaining 1%, only 0.035% is carbon dioxide.

- These percentages change very little because there is a balance between the processes that use up and make both carbon dioxide and oxygen.

- Some of these processes are shown in the **carbon cycle**. The arrows in the diagram show the direction of movement of carbon compounds.

The atmosphere

- Gases escaping from the interior of the Earth formed the original **atmosphere**.
  - Plants that could photosynthesise removed carbon dioxide from the atmosphere and added oxygen.
  - Eventually the amount of oxygen reached its current level.

Pollution control

- Pollutants are substances made by human activity that harm the environment.

- The atmosphere contains a large number of pollutants. The main ones are shown in the table.

- Scientists are finding ways to reduce the amount of these pollutants.
  - Most cars are now fitted with **catalytic converters**.
  - A catalytic converter reduces the levels of **carbon monoxide** and oxides of nitrogen.

- It is important to control atmospheric **pollution** because of the effects it can have on people's health, the natural environment and the built environment.

- Sulfur dioxide is a pollutant that can cause difficulties for people with asthma.

- A car fitted with a catalytic converter changes carbon monoxide into carbon dioxide.

Pollutant	Environmental effects
carbon monoxide	poisonous gas formed by incomplete combustion in petrol- or diesel-powered motor vehicles
oxides of nitrogen	photochemical smog and acid rain, formed by reaction of nitrogen and oxygen at very high temperatures such as those in internal combustion engines
sulfur dioxide	acid rain formed from sulfur impurities when fossil fuels burn erodes stonework, corrodes metals, kills plants and fish, and can lead to trees dying

Remember!
Combustion, respiration and photosynthesis are all vital in the carbon cycle.

Improve your grade

Evolution of the atmosphere
Describe what gases are present in the atmosphere and how the levels of these gases are maintained.
AO1 [5 marks]

Making polymers

Hydrocarbons

- The two elements carbon and hydrogen chemically combine to make a hydrocarbon.
  - **Alkanes** are hydrocarbons.
  - **Alkenes** are hydrocarbons.
- A hydrocarbon can be recognised from its **displayed formula** as it only has C or H joined together.
  - One **atom** of carbon and four atoms of hydrogen chemically combine to make a hydrocarbon called methane. Methane is an alkane.
  - Two atom of carbon and four atoms of hydrogen make a hydrocarbon called ethene which is an alkene.

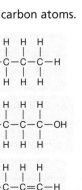

- A hydrocarbon is a **compound** of carbon and hydrogen atoms only.
  - Alkanes are hydrocarbons that have **single covalent bonds** only.
  - Alkenes are hydrocarbons that have a **double covalent bond** between a pair of carbon atoms. Double bonds involve two shared pairs of electrons.

  - Propane, C_3H_8, is a hydrocarbon because it has only C and H atoms. It is an alkane because all the bonds are single covalent bonds.

  - Propanol, C_3H_7OH, is *not* a hydrocarbon because it contains an oxygen atom.

  - Propene, C_3H_6, is a hydrocarbon because it contains only C and H atoms. It is an alkene because it has a double covalent bond between carbon atoms. Propene is also a monomer. Poly(propene) is the **polymer**.

- **Bromine** is used to test for an alkene. When orange bromine water is added to an alkene it turns colourless (**decolourises**).

Polymerisation

- A polymer is:
  - a very big molecule
  - a very long-chain molecule.
- Polymer molecules are made from many small molecules called monomers.
- When lots and lots of monomers are joined to make a polymer, the reaction is called polymerisation.

- Addition polymerisation is the process in which many alkene monomers react to give a polymer. This reaction needs high pressure and a **catalyst**.

- You can recognise a polymer from its displayed formula by looking out for the following: a long chain, the pattern repeating every two carbon atoms, two brackets on the end with extended bonds through them, an '*n*' after the brackets.

- This is the displayed formula of poly(ethene):

EXAM TIP

A polymer always has a bracket around the name of its monomer.

If ethene is the monomer then the polymer is poly(ethene).

Improve your grade

Interpreting displayed formulae

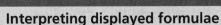

The displayed formula of butene is shown. How do you know butene is a hydrocarbon and what is the name of the polymer made from butene? *AO2* [2 marks]

Designer polymers

Breathable polymers

- Fabrics for clothes, paint for cars and cases for computers are all made from different polymers.

Polymer	Property 1	Property 2	Use
poly(vinyl chloride)	waterproof	flexible	raincoats
poly(ethene)	waterproof	flexible	plastic bags
poly(styrene)	insulates	absorbs **shock**	packaging
poly(propene)	strong	flexible	ropes
GORE-TEX®	waterproof	breathable	raincoats

- Each polymer is chosen carefully for the job that it does best.
  - Nylon and polyester are used for clothing as they can make flexible fibres.
  - Poly(propene) is used to make strong, flexible ropes.
  - One property of poly(vinyl chloride) (PVC) is that it is waterproof. Raincoats made from PVC keep the wearer dry.
  - GORE-TEX® makes raincoats that are 'breathable'.

Remember!

You need to make a reasoned judgement about why you would choose a polymer for a use. Always refer to the data given.

- Nylon is tough, lightweight, keeps water out and keeps UV (ultraviolet) light out but does not let water vapour through. This means that sweat condenses and makes the wearer wet and cold inside their jacket.

- GORE-TEX® has all the properties of nylon but is also breathable, so it is worn by many active outdoor people. Water vapour from sweat can pass through the membrane but rainwater cannot.

Disposing of polymers

- Most polymers are non-biodegradable.
  - They do not decay and are not **decomposed** by bacteria. This can cause problems.

- Some ways that waste polymers can be **disposed** of are:
  - in landfill sites
  - by burning
  - by **recycling**.

- There are problems with using non-**biodegradable** polymers.
  - They are difficult to dispose of.
  - They can cause litter.

- Scientists are developing new types of polymers:
  - polymers that dissolve
  - biodegradable polymers.

- Research into new polymers is important because there are environmental and economic issues with the use of existing polymers.
  - Disposal of non-biodegradable polymers means landfill sites get filled quickly.
  - Landfill means wasting land that could be valuable for other purposes.
  - Disposal by burning waste plastics makes **toxic** gases.
  - Disposal by burning or using landfill sites wastes the **crude oil** used to make the polymers.
  - It is difficult to sort out different polymers so recycling is difficult.

Improve your grade

Properties and uses of polymers

Explain which of these polymers would be a good material to make drain pipes, electrical cable covers and socks. *AO2* [3 marks]

Property	A	B	C
waterproof	no	yes	yes
electrical conductor	no	no	no
flexible	yes	no	yes

Cooking and food additives

The changing of food

- A chemical change takes place if:
  - a new substance is made
  - the change is irreversible
  - an **energy** change takes place.
- These chemical changes happen when the chemicals in food are heated.
  - When food is cooked a new substance is made.
  - The process cannot be reversed.
- Protein molecules in eggs and meat permanently change shape when eggs and meat are cooked.
- This changing of shape is called 'denaturing'.

Proteins denature on heating

G–E

D–C

Baking powder

- Baking powder is added to flour to make cakes rise.
  - **Carbon dioxide** is made when baking powder is heated in an oven.
  - The carbon dioxide makes the cake rise.
- Baking powder is sodium hydrogencarbonate.
- When it is heated it breaks down (**decomposes**) to give carbon dioxide.
- The word equation for the decomposition of sodium hydrogencarbonate is:

  sodium hydrogencarbonate → sodium carbonate + carbon dioxide + water

- The **balanced symbol equation** for the decomposition of sodium hydrogencarbonate is:

  $$2NaHCO_3 \rightarrow Na_2CO_3 + CO_2 + H_2O$$

Remember!

When carbon dioxide is bubbled through limewater, the colourless liquid turns cloudy.

EXAM TIP

In an exam question which asks you to construct the word equation for the decomposition of sodium hydrogencarbonate, not all the products will be given to you.

G–E

D–C

Additives and emulsifiers

- When foods are processed, additives are often put in for different purposes.
- The main types of food additive are:
  - antioxidants, which stop food from reacting with oxygen
  - food colours, which give a better colour
  - flavour enhancers, which improve the flavour of food
  - emulsifiers, which help oil and water to mix and not separate.
- Emulsifiers are additives that help oil and water mix in foods.
  - Mayonnaise is made by mixing oil and vinegar with egg. Egg is the emulsifier. It is needed because oil does not attract water. Oil repels water.
- Emulsifiers are molecules that have a water-loving (hydrophilic) part and an oil- or fat-loving (hydrophobic) part.
- The oil- or fat-loving part (the hydrophobic end) goes into the fat droplet.

hydrophilic head

hydrophobic tail

An emulsifying molecule

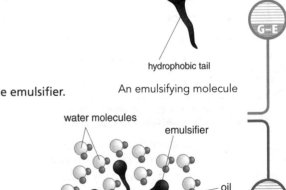

water molecules

emulsifier

oil drop

An emulsion of oil and water

G–E

D–C

Improve your grade

Using baking powder

How does baking powder, sodium hydrogencarbonate, help cakes to rise?
Write a word equation for the reaction that happens. *AO1* [3 marks]

Smells

Esters

G–E

- Some perfumes and cosmetics are made from natural sources. Others can be made synthetically.
- Oil from roses can be **distilled** to make perfume. Lavender oil is made from lavender grown in Norfolk. These are perfumes from natural sources.
- Similar perfumes can be made synthetically. Chemicals are boiled to make an ester that has the smell needed.

D–C

- **Alcohols** react with **acids** to make an ester and water.

  alcohol + acid → ester + water

- Esters are used to make perfumes.
- An ester can be made using a simple experiment.
  – The acid is added to the alcohol and heated for some time in a water bath.

Perfume properties

G–E

- A good perfume has a pleasant smell. It also needs the properties shown in the diagram.

D–C

- Perfumes need these properties because they must:
  – **evaporate** easily so that the perfume particles can reach the nose
  – be non-toxic so it does not poison you
  – not react with water so the perfume does not react with perspiration
  – not irritate the skin so the perfume can be put directly on to the skin
  – be insoluble in water so it cannot be washed off easily.

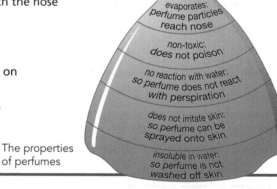

evaporates: perfume particles reach nose

non-toxic: does not poison

no reaction with water: so perfume does not react with perspiration

does not irritate skin: so perfume can be sprayed onto skin

insoluble in water: so perfume is not washed off skin

The properties of perfumes

Solubility

G–E

- A substance that dissolves in a liquid is **soluble**.
  – The substance dissolved in a solution is called the solute.
  – The liquid that it dissolves in is called the solvent.
  – A substance that does not dissolve in a liquid is insoluble.
- Water does not dissolve nail varnish so it cannot be used to remove varnish from nails.
- However, nail varnish remover does dissolve nail varnish.
- A solution is a mixture of solvent and solute that does not separate out.

D–C

- Esters can be used as solvents.

> **Remember!**
> You will need to interpret data on the effectiveness of solvents.

Cosmetics testing

G–E

- All cosmetics must be tested to make sure they are safe and do not harm humans.
  – They must not cause rashes or itchiness.
  – They must not cause skin damage or lead to cancer or other life-threatening conditions in long-term use.
- The EU has banned all testing of cosmetics on animals.

How science works

You should be able to:
- explain why testing of cosmetics on animals has been banned in the EU.

Improve your grade

Esters and perfumes

Explain what specific properties a perfume needs as well as a pleasant smell. *AO1* [3 marks]

C1 Carbon chemistry

Paints and pigments

Colloids

- The ingredients of paint are:

pigment	the substance that gives the paint its colour
binding medium	sticks the pigment in the paint to the surface
solvent	the solvent thins the paint making it easier to use

- In oil paints:
  - The pigment powder is dispersed (spread) through the oil.
  - A solvent is often added that dissolves oil.

- We use paint for two main reasons:
  - for protection – woodwork outside is painted to protect the wood against rain; the oil sticks to the wood and forms a skin
  - to look attractive – when pictures or walls inside a house are painted, it is the pigment part of the paint we look at; the binder is used to stick pigment to the canvas or walls.

- Paint is a **colloid** where the particles are mixed and dispersed with particles of a liquid (binding medium) but are not dissolved.

G–E

D–C

Paint drying

- Most paints dry because:
  - paints are applied as a thin layer
  - the solvent evaporates.

- Emulsion paints are water-based paints that dry when the solvent evaporates.

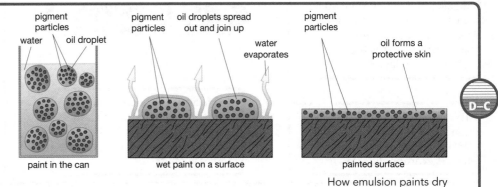

paint in the can

wet paint on a surface

painted surface

How emulsion paints dry

D–C

Thermochromic pigments

- The pigments dissolved in substances to make thermochromic paints are very unusual. When they get hot they change colour. When they cool down they change back to their original colour.

- Thermochromic pigments can be used in cups to change colour when the cup is hot. They can provide a warning if a kettle is getting hot. They can be used as bath toys to change colour in warm water.

- Thermochromic pigments used in some paints are chosen for their colour and also for the **temperature** at which their colour changes.
  - Many people find that anything over 60 °C is too hot to hold, so a thermochromic pigment that changes colour at 45 °C can be used to paint cups or kettles to act as a warning.
  - A pigment that changes colour just above 0 °C makes a good warning paint for road signs to show if the road might freeze.

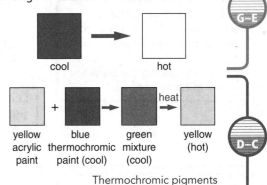

cool   hot

yellow acrylic paint + blue thermochromic paint (cool) → green mixture (cool) → heat → yellow (hot)

Thermochromic pigments

G–E

D–C

Phosphorescent pigments

- A phosphorescent pigment glows in the dark.

- Phosphorescent pigments glow in the dark because:
  - they absorb and store **energy**
  - they release the energy as light over a period of time.

G–E

D–C

Improve your grade

Drying paint

Explain what paint is made up of and how phosphorescent pigments work. *AO1* [3 marks]

C1 Summary

Crude oil is a mixture of many hydrocarbons.

Fossil fuels are finite resources because they are no longer being made or are being made extremely slowly. They are non-renewable resources as they are being used up faster than they are being formed.

Making crude oil useful

Crude oil can be separated by fractional distillation. It is heated near the bottom of a fractionating column. Fractions with low boiling points 'exit' at the top. Fractions with high boiling points exit at the bottom. There is a temperature gradient between the bottom and the top of the column.

Cracking is a process which makes more petrol, using a catalyst and high temperatures. Cracking converts large alkane molecules into smaller alkane and alkene molecules.

Fractional distillation works because the fractions of crude oil have different boiling points. The fractions are LPG, petrol, diesel, paraffin, heating oil, fuel oils and bitumen.

If hydrocarbons burn in a plentiful supply of air, carbon dioxide and water are made. In an experiment, carbon dioxide can be tested for with limewater; it turns the limewater milky.

Fuels and clean air

The original atmosphere came from gases escaping from the interior of the Earth. Photosynthesis by plants increased the percentage of oxygen to the current level.

The present day atmosphere contains 21% oxygen, 78% nitrogen and 0.035% carbon dioxide.

Key factors that need to be considered when choosing a fuel are energy value, availability, storage, toxicity, pollution caused and ease of use.

Photosynthesis, respiration and combustion are processes in the carbon cycle. Photosynthesis increases the level of oxygen, respiration and combustion decrease the level of oxygen.

A hydrocarbon is a compound of carbon atoms and hydrogen atoms only.

Addition polymers are made when alkene monomer molecules react together under high pressure and with a catalyst.

Nylon and GORE-TEX® are polymers with suitable properties for particular uses. Nylon is tough, lightweight and keeps water and UV light out. GORE-TEX® has all these properties but allows water vapour to pass out so that sweat does not condense.

Alkanes are hydrocarbons which contain single covalent bonds only. Alkenes are hydrocarbons which contain a double covalent bond between a pair of carbon atoms.

Polymers

Many polymers are non-biodegradable and so will not decompose by bacterial action. They are disposed of in landfill sites or by burning, or are recycled.

Protein molecules in eggs and meat denature when they are cooked. The change of food during cooking is a chemical change as it is an irreversible process and a new substance is made.

Food, smells and paint

Paints are colloids because the particles are mixed and dispersed throughout a liquid but are not dissolved in it. Thermochromic pigments change colour with heat and are used, for example, in babies' toys that are immersed in water to show whether the water is too hot. Phosphorescent pigments glow in the dark as they absorb and store energy then release it as light over a period of time.

Emulsifiers help oil and water to mix and not to separate. They are molecules with a water-loving (hydrophilic) part and an oil-loving (hydrophobic) part.

Perfumes need to be able to evaporate so that they can easily reach the nose. They need to be non-toxic, insoluble in water, not irritate the skin, and not react with water. Perfumes are esters which are often made synthetically.

The structure of the Earth

What is the Earth made from?

- The Earth is a sphere with an iron **core**, surrounded by a **mantle**.
- The outer **crust** is thin and made of rock.

thin rocky crust

mantle

core with iron in it

The structure of the Earth

More on the Earth's structure

- The outer layer of the Earth is called the **lithosphere**. This layer is (relatively) cold and rigid and comprises the crust and top part of the mantle.
- The lithosphere is made of **tectonic plates** which are less dense than the mantle below.
- The crust is too thick to drill through, so most of our information about the Earth is collected from **seismic waves** produced by earthquakes and man-made **explosions**.

Does the Earth move?

- The outer layer of the Earth is made from tectonic plates.
- Plate movement causes volcanoes and earthquakes at the plate boundaries.
- Tectonic plates move very slowly (about 2.5 cm a year). This means continents change position over millions of years.
- Different theories have been suggested about why the Earth's surface is like it is.
- Most Earth scientists now accept the theory of plate tectonics, which suggests that Africa and South America were once one land mass.
- The theory is accepted, because:
  - it explains a wide range of evidence
  - it has been discussed and tested by many scientists.

What happens to molten rock?

- Molten rock beneath the Earth's surface is called **magma**. Molten rock at the Earth's surface is called lava.
- When volcanoes erupt, the lava can either be:
  - runny, slow moving and fairly safe, or
  - thick and erupt explosively, causing major damage.
- People live near volcanoes because the ash is good for growing crops.
- When molten rock cools down, it makes **igneous rock**. This contains crystals which form in different sizes.
  - If the rocks cool slowly underground, large crystals form.
  - If the rocks are close to the surface, they cool quickly, and small crystals form.
- Magma rises up through the Earth's crust because it is less dense than the crust. This can cause volcanoes.
- Magma can have different types of composition which cause different types of eruption.
- Geologists study volcanoes to try to forecast future eruptions and reveal more about the structure of the Earth.

How science works

You should be able to:
- describe a simple scientific idea using a model
- recognise that scientific explanations are provisional but more convincing when there is more evidence to support them.

Improve your grade

Living near a volcano

When a volcano erupts, it releases lava. What affects the type of volcanic eruption? *AO1* [3 marks]

Construction materials

Raw materials for building

G–E

- **Granite**, **marble** and **limestone** are rocks that are used to make buildings.
- Limestone and aggregate are used to make roads.
- Mining or quarrying can be used to get rocks out of the ground.
- Mining and quarrying:
  - cause environmental problems such as noise, dust and increased traffic
  - destroy the landscape, which needs to be reconstructed when the quarrying and mining is finished.

D–C

- Some raw materials used to make construction materials are found in the Earth's **crust**.

Raw material	clay	limestone and clay	sand	iron ore	aluminium ore
	↓	↓	↓	↓	↓
Building material	brick	cement	glass	iron	aluminium

- Hardness can be compared by rubbing two materials together. Granite is harder than marble, and marble is harder than limestone.
- Many buildings look like they are made from one material, but they are often made with different materials and only lined with an expensive material.

> **EXAM TIP**
>
> In questions about using materials, try to use a different property for each answer.

Cement and concrete

G–E

- Limestone and marble are both forms of calcium carbonate.
- When limestone is heated, it breaks down into calcium oxide and **carbon dioxide**. This is called **thermal decomposition**.
- Limestone is used to make **cement**.
- **Concrete** is made when cement, sand, small stones (aggregate) and water are mixed and allowed to set.
- Concrete can be made stronger by letting it set around a grid of steel or steel rods.

D–C

- The thermal decomposition of calcium carbonate can be represented by the equation:

$$\text{calcium carbonate} \rightarrow \text{calcium oxide} + \text{carbon dioxide}$$
$$CaCO_3 \rightarrow CaO + CO_2$$

- Cement is made when limestone is heated with clay.
- **Reinforced concrete** is a **composite material** which has steel rods or meshes running through it. Composites contain at least two materials that can still be distinguished.

> **Remember!**
>
> Thermal decomposition is a reaction where one substance breaks down on heating to give at least two new substances.

Improve your grade

Rock hardness

Describe how concrete can be made. *AO1* [2 marks]

Metals and alloys

Making and extracting copper

- Rocks containing copper ore are mined. Copper is a metal **element**; ores are **compounds**.

- Copper can be extracted from its ore by heating the ore with carbon to remove the oxygen.

- The removal of oxygen from a substance is called **reduction**.

- **Recycling** copper is preferable to extracting it from its ore because:
  - it saves valuable copper ore resources
  - it saves the energy needed to crush rocks and operate machinery.

- Copper can be purified by **electrolysis.**

- Impure copper can be purified in the laboratory using an electrolysis cell.

> **Remember!**
> Copper is extracted from its ore by reduction, but purified by electrolysis.

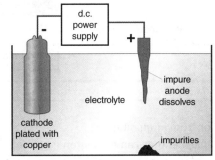

Purifying copper by electrolysis

- Advantages of recycling copper are that:
  - it has a fairly low **melting point** so the energy cost to melt it is low
  - it reduces the need for mining, saving reserves and the environmental problems caused by mining
  - it keeps the cost of copper down.

- Disadvantages and problems of recycling copper are that:
  - the small amounts used in electrical equipment are difficult to separate
  - valuable 'pure' copper scrap must not be mixed with less pure scrap, such as **solder**
  - less copper is mined so there are fewer mining jobs
  - the actual separating process may produce **pollution**
  - a lot of copper is thrown away as it is difficult to persuade people to recycle it.

Alloys

- **Alloys** are mixtures containing one or more metal elements.

- Examples of alloys are **brass**, bronze, solder, steel and **amalgam**.

- By adding another element to a metal, its properties change, making it more useful. For example, steel is a useful alloy of iron.

- Important large-scale uses of alloys are:
  - amalgam used in tooth filings
  - brass used for door hinges, musical instruments and to make coins
  - solder used to join electrical wires.

- The main metals in each of these alloys are:
  - amalgam – mercury
  - brass – copper and zinc
  - solder – **lead** and tin.

EXAM TIP
You need to be able to interpret data about metals and alloys, e.g. hardness, density, boiling point and strength, and suggest properties for different uses.

Improve your grade

Uses of metals
Three metals have properties on a relative scale where 1 is the least and 10 is the most.

Using the information in the table, explain why you would use aluminium for aircraft bodies, iron to construct bridges and lead for protecting joins in roofs. *AO2* [3 marks]

Metal	Density	Strength	Hardness
aluminium	1	5	4
iron	6	8	9
lead	10	1	1

Making cars

Rusting and corrosion

- **Rust** is a brownish solid which forms when iron is in contact with water and oxygen.
- Rust forms slowly and flakes off from the surface. Rusting is an example of **oxidation** – adding oxygen to a substance.
- Aluminium does not **corrode** in damp conditions.

- Only iron and steel rust. Other metals corrode.
- **Acid rain** and **salt** water accelerate rusting.
- Rusting is an oxidation reaction because iron reacts with oxygen forming an oxide.
- The word equation for rusting is:

  iron + oxygen + water → **hydrated iron(III) oxide**

- Aluminium does not corrode in moist air because it has a protective layer of aluminium oxide which, unlike rust, does not flake off the surface.
- Different metals corrode at different rates.

> ### EXAM TIP
> Word equations must be given in words. If you use symbols you may not get the marks.

Materials used in cars

- Car bodies can be made from steel (mostly iron) or from aluminium.
  - Both are good **electrical conductors** and **malleable**, i.e. they can be beaten into shape.
  - In contrast to iron, aluminium has a low **density**, does not corrode and is not magnetic.
- The main metals used to build a car are steel, copper and aluminium.
- The main **non-metals** used to build a car are glass, plastic and fibres.

- Different materials are used in cars because they have different properties.
  - Copper is used for electrical wire as it is a good conductor.
  - Plastic/glass windscreens are transparent.
  - Plastic dashboards are cheap and rigid.
  - Textiles in seats are hardwearing, but soft.
- **Alloys** are mixtures of **elements** containing at least one metal, and often have different and more useful properties than the metals they are made from. For example:
  - steel is harder and stronger than iron
  - steel is less likely to corrode than iron.
- Car bodies can be built from aluminium or steel and there are advantages and disadvantages with each. Aluminium is lighter and more resistant to corrosion than steel. However, steel costs less and is stronger.

Recycling

- There are advantages to recycling materials – it:
  - saves natural resources such as ores
  - reduces the amount of rubbish that needs to be got rid of.

- Advantages of **recycling** the materials from old cars include:
  - less mining saves **finite resources** needed to make metals
  - less **crude oil** is needed to make new plastics
  - less waste means less landfill
  - fewer **toxic** materials, such as **lead** from **batteries**, are dumped.
- Disadvantages of recycling the materials from old cars include:
  - fewer mines are built and fewer mining jobs created
  - difficult to separate the different materials
  - some separating techniques produce **pollution**
  - some recycling processes are very expensive.
- There are laws which specify that a minimum percentage of all materials used to manufacture cars must be recyclable to help protect the environment.

Improve your grade

Recycling cars
Suggest why people recycle materials from old cars. *AO1* [2 marks]

Manufacturing chemicals – making ammonia

The Haber process

- Ammonia is a gaseous compound containing one **atom** of nitrogen and three atoms of hydrogen.

- Ammonia is made using the **Haber process**.

- The Haber process uses:
  - nitrogen from the air
  - hydrogen from natural gas or the **cracking** of oil fractions.

- Making ammonia is a reversible reaction.

- The ⇌ symbol is used to represent a reversible reaction.

- Reversible reactions can go in both directions. As ammonia is made, it breaks back down into nitrogen and hydrogen.

- Two of the products made from ammonia are:
  - **fertilisers**
  - nitric acid.

The Haber process

- World food production depends on nitrogen fertilisers. These fertilisers are made from ammonia, which is made by the Haber process.
  - The word equation for the Haber process equation is:

    nitrogen + hydrogen ⇌ ammonia

  - The **balanced symbol equation** for the Haber process equation is:

    $N_2 + 3H_2 \rightleftharpoons 2NH_3$

- The **optimum** (best) **conditions** for the Haber process are to:
  - use a **catalyst** made of iron
  - raise the **temperature** to about 450 °C (fairly low for an industrial process)
  - use high pressure (about 200 atmospheres/20 MPa)
  - **recycle** any unreacted nitrogen and hydrogen.

What affects the cost of chemical manufacture?

- The cost of making a new substance depends on:
  - the price of energy (gas and electricity)
  - the cost of the starting materials
  - people's wages
  - the cost of buying equipment (plant)
  - how quickly the new substance can be made (including the cost of the catalyst).

- Costs increase when the pressure is raised (increasing the plant costs), and the temperature is raised (increasing the energy costs).

- Costs decrease when catalysts are used, unreacted starting materials are recycled and automation is used (reducing wage bills).

EXAM TIP

Make sure you know how ammonia is made, and what it is used for.

G–E

D–C

G–E

D–C

Improve your grade

The manufacture of ammonia

Describe how ammonia is manufactured by the Haber process. *AO1* [4 marks]

Acids and bases

Acids, bases, alkalis and neutralisation

- Universal indicator can be used to estimate the **pH** of a solution.

- Universal indicator changes colour at different pH levels:
  - pH less than 7 shows the solution is acidic (red is strong, orange/yellow is weak)
  - pH 7 shows the solution is neutral (green)
  - pH above 7 shows the solution is alkaline (deep blue/purple is strong, turquoise is weak).

| (low pH) ACID pH = 0 acids have a pH of less than 7 | ← | NEUTRAL pH = 7 | → | (high pH) ALKALI pH = 14 alkalis have a pH of more than 7 |

The pH scale

- Litmus is another common indicator:
  - **Acids** turns blue litmus red.
  - **Alkalis** turn red litmus blue.

- An acid can be **neutralised** by adding an alkali or **base** and vice versa.

- An alkali is a **soluble** base.

- If an alkali is added to an acid, the pH number will gradually *increase*. When it gets to pH 7 the acid is neutralised.

- If acid is added to an alkali, the pH number will gradually *decrease*. When it gets to 7 the alkali has been neutralised.

- Metal oxides and metal hydroxides are bases. A few bases are soluble in water, and are called alkalis, e.g. sodium hydroxide and calcium hydroxide.

- *Neutralisation* takes place when an acid and a base react to make **salt** and water. The word equation for neutralisation is:

  acid + base → salt + water

Remember!
Acids react with alkalis to make salt and water. This process is called neutralisation.

- Some indicators show a sudden colour change at one **pH** value. Universal indicator shows a gradual range of colour changes, as it contains a mixture of different indicators.

- In **solution**, all acids contain H^+ **ions** (hydrogen ions). The pH of an acid is determined by the **concentration** of H^+ ions – the higher the concentration the lower the pH. Neutralisation leaves no free H^+ ions.

Salts

- Acids react with bases and metal carbonates to form salts.

- The word equations for making salt are:

  acid + base → salt + water

  acid + metal carbonate → salt + water + carbon dioxide

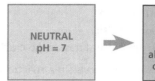
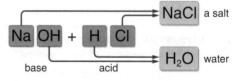

The reaction of sodium hydroxide with hydrochloric acid produces sodium chloride

- Salt names have two parts – the second part of the name shows which acid it has been made from. For example:

Common acid	Second part of the salt name
sulfuric	sulfate
nitric	nitrate
hydrochloric	chloride
phosphoric	phosphate

Improve your grade

Estimating pH

Each of four solutions – A, B, C and D – have a different pH.
Explain how you can estimate whether these pH values are correct and what happens if B and D are mixed. *AO2* [3 marks]

Solution	A	B	C	D
pH	2	6	7	9

C2 Chemical resources

Fertilisers and crop yield

Growing crops

- **Fertilisers** increase crop yield. They contain essential chemical elements in the form of minerals.

- Plants absorb these minerals though their roots. This means that the minerals need to be dissolved in water before the roots can absorb them.

- Three **essential elements** needed for good plant growth are nitrogen (N), phosphorus (P) and potassium (K).

- Fertilisers containing all three of these elements are often called NPK fertilisers.

- NPK values are percentages, and are always listed in order.
  So a 12–8–10 fertiliser has 12% nitrogen, 8% phosphorous and 10% potassium.

- The formula of a fertiliser shows the essential elements it contains. For example, KNO_3 contains potassium (K) and nitrogen (N).

G–E

- Fertilisers must be dissolved in water before they can be absorbed by the plant roots as only dissolved substances are small enough to be absorbed.

- Some fertilisers dissolve easily, but some are designed for slow release giving crops a small amount over a long time.

- Fertilisers are needed because the world **population** is rising and there is a greater demand for food production from the land available.

Remember!
Fertilisers help food production for a rising population, but excessive use can cause eutrophication.

D–C

Eutrophication

- Fertilisers are very useful because they increase crop yields, such as producing more and larger grains of wheat, increasing the food supply.

G–E

- However, fertilisers can also cause problems. If they are washed into rivers and lakes as they can cause the death of animals that live in the water (this process is called **eutrophication**).

- Eutrophication and pollution of water supplies can result from the excessive use of fertilisers.

D–C

Preparing fertilisers

- Fertilisers are made by **neutralising** an **acid** with an **alkali**.

- Fertilisers that contain nitrogen are called *nitrogenous fertilisers* and are manufactured from ammonia. Some examples are:
  - ammonium nitrate
  - ammonium sulfate
  - ammonium phosphate
  - urea.

- It is possible to predict the acid and alkali used to produce a fertiliser from its name. For example, ammonium nitrate is made by combining nitric acid and ammonia solution.

1. Use a measuring cylinder to pour alkali into a conical flask.

measuring cylinder

conical flask

2. Add acid to the alkali until it is neutral.

burette

G–E

3. Evaporate.

evaporating basin

crystals begin to form

4. Filter off the crystals.

filter paper

filter funnel

D–C

Making a fertiliser

Improve your grade

Essential elements in a fertiliser
What do fertilisers contain and how do they get into the plant? *AO1* [3 marks]

Salt of the Earth

- Sodium chloride (**salt**) is an important raw material.
- Salt can be obtained by mining salt buried in the Earth or by evaporating **sea water**.
- Salt is mined in Cheshire in two different ways:
  - mining it from the ground as rock salt
  - solution mining by pumping in water and extracting saturated salt **solution**.
- Mining salt can lead to **subsidence**. The ground above a mine can sink causing landslips and destroying homes.
- Salt at the surface, particularly brine solution, can escape and affect **habitats**.

Electrolysis of sodium chloride solution

- Passing electricity through a solution is called **electrolysis**.
- The electrolysis of concentrated sodium chloride solution makes chlorine gas and hydrogen gas.
- Chlorine bleaches damp litmus paper. This is the chemical test for chlorine.
- During the electrolysis of concentrated sodium chloride solution (brine):
  - hydrogen is made at the negative **cathode**
  - chlorine is made at the positive **anode**
  - sodium hydroxide forms in solution.
- Hydrogen and chlorine are reactive, so it is important to use inert electrodes so that the products don't react before they are collected and the electrodes do not dissolve.

hydrogen gas out | brine in | chlorine gas out

sodium hydroxide solution

H_2 | Cl_2

cathode | porous barrier | anode

Products of electrolysis

The chlor-alkali industry

- Sodium chloride is used in food as:
  - a preservative
  - a flavouring.
- Sodium chloride is a very important raw material in the chemical industry.
- Sodium chloride solution is split up to make:
  - chlorine for use in sterilising water supplies, for making household bleach, plastics such as PVC, and solvents to dissolve other chemicals
  - hydrogen for use in margarine production
  - sodium hydroxide for use in soap-making.
- Sodium hydroxide and chlorine are used to make household bleach.

EXAM TIP

Sodium chloride solution is split up by electrolysis into hydrogen, chlorine and sodium hydroxide. Learn a use for each one.

Improve your grade

Splitting up sodium chloride

How can concentrated sodium chloride (salt) solution be split up, and what are the products of the reaction? *AO1* [2 marks]

C2 Summary

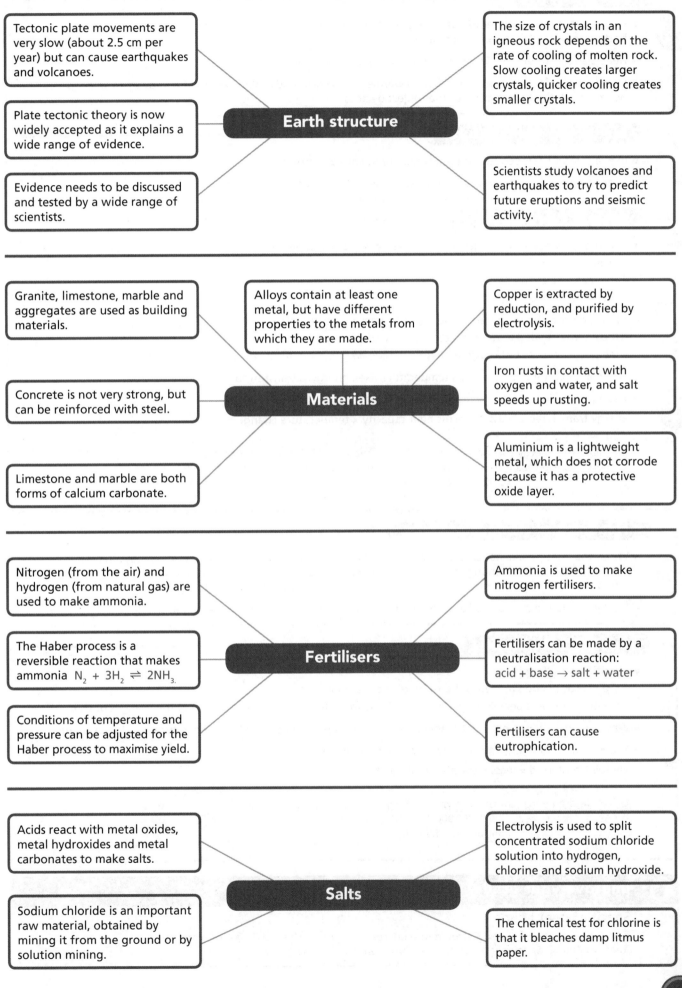

Earth structure

Tectonic plate movements are very slow (about 2.5 cm per year) but can cause earthquakes and volcanoes.

Plate tectonic theory is now widely accepted as it explains a wide range of evidence.

Evidence needs to be discussed and tested by a wide range of scientists.

The size of crystals in an igneous rock depends on the rate of cooling of molten rock. Slow cooling creates larger crystals, quicker cooling creates smaller crystals.

Scientists study volcanoes and earthquakes to try to predict future eruptions and seismic activity.

Materials

Granite, limestone, marble and aggregates are used as building materials.

Alloys contain at least one metal, but have different properties to the metals from which they are made.

Concrete is not very strong, but can be reinforced with steel.

Limestone and marble are both forms of calcium carbonate.

Copper is extracted by reduction, and purified by electrolysis.

Iron rusts in contact with oxygen and water, and salt speeds up rusting.

Aluminium is a lightweight metal, which does not corrode because it has a protective oxide layer.

Fertilisers

Nitrogen (from the air) and hydrogen (from natural gas) are used to make ammonia.

The Haber process is a reversible reaction that makes ammonia $N_2 + 3H_2 \rightleftharpoons 2NH_3$.

Conditions of temperature and pressure can be adjusted for the Haber process to maximise yield.

Ammonia is used to make nitrogen fertilisers.

Fertilisers can be made by a neutralisation reaction:
acid + base → salt + water

Fertilisers can cause eutrophication.

Salts

Acids react with metal oxides, metal hydroxides and metal carbonates to make salts.

Sodium chloride is an important raw material, obtained by mining it from the ground or by solution mining.

Electrolysis is used to split concentrated sodium chloride solution into hydrogen, chlorine and sodium hydroxide.

The chemical test for chlorine is that it bleaches damp litmus paper.

Heating houses

Energy flow

G–E

- Hot objects have a high **temperature** and usually cool down. The rate of cooling depends on how much hotter the body is than its surroundings.

- Heat is a form of **energy**. The unit of energy (heat) is the **joule** (J).

D–C

- Energy, in the form of heat, flows from a warmer to a colder body. When energy flows away from a warm object, the temperature of that object decreases.

Measuring temperature

G–E

- In a laboratory, temperature is usually measured with a thermometer.

- **Thermograms** are pictures in which colour is used to represent temperature.

Specific heat capacity

G–E

- The energy needed to change the temperature of a body depends on:
  - its **mass**
  - the material it is made from
  - the temperature change.

D–C

- All substances have a property called **specific heat capacity**, which is:
  - the energy needed to raise the temperature of 1 kg by 1 °C
  - measured in joule per kilogram degree Celsius (J/kg °C) and differs for different materials.

- When an object is heated and its temperature rises, energy is transferred.

- The equation for energy transfer by specific heat capacity is:

energy transferred = mass × specific heat capacity × temperature change

> Calculate the energy transferred when 30 kg of water cools from 25 °C to 5 °C.
> energy transferred = 30 × 4200 × (25 − 5) i.e. 30 × 4200 × 20
> = 2 520 000 J or 2520 kJ

Heating and changes of state

G–E

- When you heat a solid, its temperature rises until it changes to a liquid. The temperature stays the same until all of the solid has changed to a liquid. The temperature of the liquid then rises until it changes into a gas. The temperature stays the same until all of the liquid has changed to a gas.

- Heat is needed to melt a solid at its melting point or to boil a liquid at its **boiling point**.

Specific latent heat

D–C

- **Specific latent heat** is:
  - the energy needed to melt or boil 1 kg of the material
  - measured in joule per kilogram (J/kg) and differs for different materials and each of the changes of state.

- When an object is heated and it changes state, energy is transferred, but the temperature remains constant.

- The equation for energy transfer by specific latent heat is:

energy transferred = mass × specific latent heat

> Calculate the energy transferred when 2.5 kg of water changes from solid to liquid at 0 °C
> energy transferred = 2.5 × 340 000
> = 850 000 J or 850 kJ

Improve your grade

Specific latent heat

Julie boils a kettle of water. Calculate the energy required to change 0.2 kg of water to steam at 100 °C. Specific latent heat = 2 260 000 J/kg. *AO1/AO2* [2 marks]

Keeping homes warm

Energy transfer

- Air is a good **insulator** which reduces **energy** loss from a home so insulation materials contain a lot of trapped air.
- Solids are good conductors because their particles are close together. They can transfer energy very easily.
- **Infrared radiation** from the Sun can be reflected by a shiny surface and absorbed by a dark matt surface. The heat is used for cooking or producing electricity.

G–E

Saving energy

- Energy loss from uninsulated homes can be reduced by:
  - injecting foam into the cavity between the inner and outer walls
  - laying fibreglass or similar material between the joists in the loft
  - replacing single glazed windows with double (or triple) glazed windows
  - drawing the curtains
  - placing shiny foil behind radiators
  - sealing gaps around doors.

G–E

Practical insulation

- Double glazing reduces energy loss by conduction. The gap between the two pieces of glass is filled with a gas or contains a **vacuum.**
  - Particles in a gas are far apart. It is very difficult to transfer energy. There are no particles in a vacuum so it is impossible to transfer energy by conduction.

space filled with air or argon, or has a vacuum

A double glazed window

- Loft insulation reduces energy loss by conduction and convection:
  - warm air in the home rises
  - energy is transferred through the ceiling by conduction
  - air in the loft is warmed by the top of the ceiling
  - the warm air is trapped in the loft insulation
  - both sides of the ceiling are at the same **temperature** so no energy is transferred
  - without loft insulation, the warm air in the loft can move by convection and heat the roof tiles
  - energy is transferred to the outside by conduction.

- Cavity wall insulation reduces energy loss by conduction and convection:
  - the air in the foam is a good insulator
  - the air cannot move by convection because it is trapped in the foam.

- Insulation blocks used to build new homes have shiny foil on both sides to reduce energy transfer by radiation:
  - energy from the Sun is reflected back to keep the home cool in summer
  - energy from the home is reflected back to keep the home warm in winter.

D–C

roof 25%

windows 10%

walls 35%

doors 15%

floors 15%

Energy loss from a home

> **Remember!**
> Hot air will only rise into the loft if the loft-hatch is open.

Energy efficiency

$$\text{efficiency} = \frac{\text{useful energy output} (\times 100\%)}{\text{total energy input}}$$

- Energy transformations can be shown by Sankey diagrams.
- Energy from the source (home) is lost to the sink (environment).
- Different types of insulation cost different amounts and save different amounts of energy.

$$\text{payback time} = \frac{\text{cost of insulation}}{\text{annual saving}}$$

100 J in coal

25 J to room

Sankey diagram

75 J to surroundings

D–C

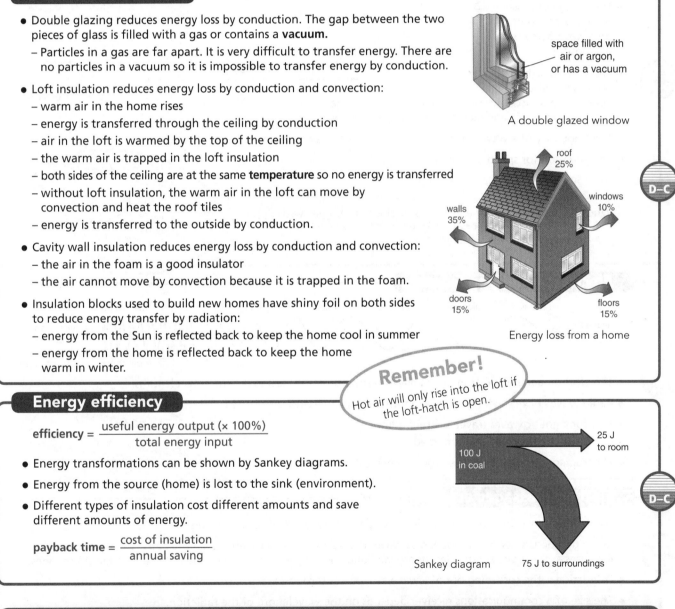

Improve your grade

Cavity wall insulation payback time

(a) The Kelly family spend £250 on cavity wall insulation. The payback time is 2 years.
    What is meant by payback time? *AO1* [1 mark]

(b) Explain how cavity wall insulation reduces energy transfer. *AO1* [3 marks]

A spectrum of waves

Transverse waves

G–E

- When a pebble is dropped into a pool of water, a circular **wave** spreads out:
  – The water particles move up and down as the wave spreads out.
- Water waves are **transverse** waves. Transverse waves travel at right angles to the wave vibration.
- Light is an example of a transverse wave and travels at 300 000 km/s.
- All waves in the **electromagnetic spectrum** travel at the **speed** of light.

Wave properties

D–C

- The **amplitude** of a wave is the *maximum* displacement of a particle *from* its rest position.
- The crest of a wave is the *highest point on* a wave *above* its rest position.
- The trough of a wave is the *lowest point on* a wave *below* its rest position.
- The **wavelength** of a wave is the distance *between* two *successive* points on a wave having the same displacement and moving in the same direction.
- The **frequency** of a wave is the number of complete waves passing a point in one second.
- The equation for speed of a wave is:

wave speed = frequency × wavelength

When Katie throws a stone into a pond, the distance between ripples is 0.3 m and four waves reach the edge of the pond each second.
wave speed = 0.3 × 4 = 1.2 m/s

Remember!
Always give the units in your answer:
– wavelength – metre (m)
– frequency – **hertz** (Hz)
– speed – metre per second (m/s).

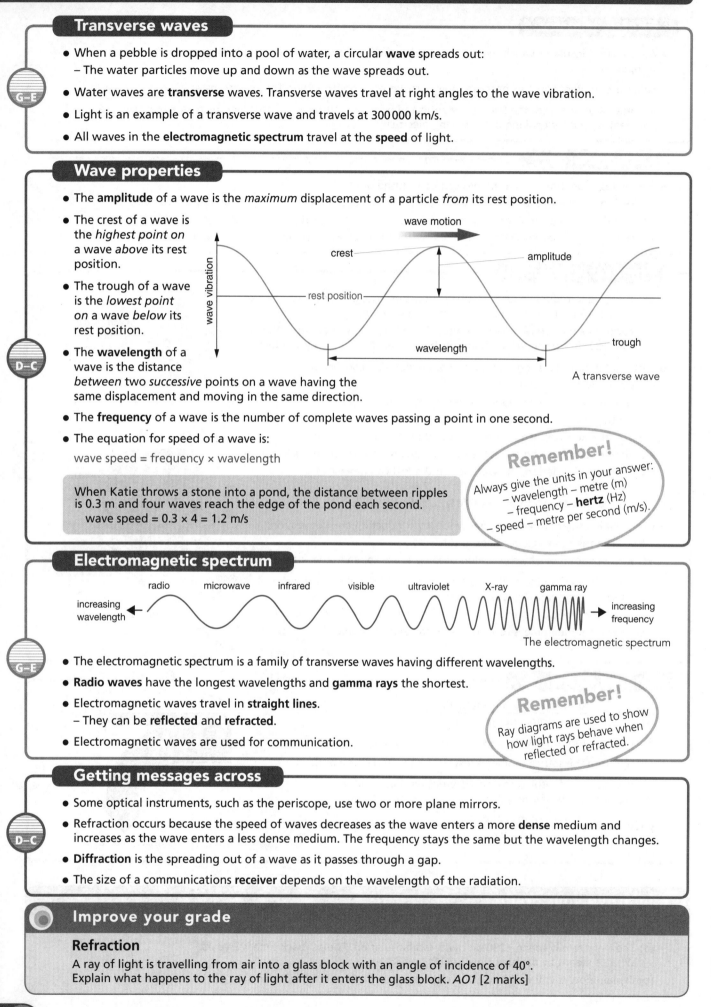

A transverse wave

Electromagnetic spectrum

The electromagnetic spectrum

G–E

- The electromagnetic spectrum is a family of transverse waves having different wavelengths.
- **Radio waves** have the longest wavelengths and **gamma rays** the shortest.
- Electromagnetic waves travel in **straight lines**.
  – They can be **reflected** and **refracted**.
- Electromagnetic waves are used for communication.

Remember!
Ray diagrams are used to show how light rays behave when reflected or refracted.

Getting messages across

D–C

- Some optical instruments, such as the periscope, use two or more plane mirrors.
- Refraction occurs because the speed of waves decreases as the wave enters a more **dense** medium and increases as the wave enters a less dense medium. The frequency stays the same but the wavelength changes.
- **Diffraction** is the spreading out of a wave as it passes through a gap.
- The size of a communications **receiver** depends on the wavelength of the radiation.

Improve your grade

Refraction
A ray of light is travelling from air into a glass block with an angle of incidence of 40°.
Explain what happens to the ray of light after it enters the glass block. *AO1* [2 marks]

Light and lasers

Sending signals

- Early messages relied on *line of sight* – smoke signals, beacons, semaphore.
- Runners and horsemen relayed messages over longer distances.
- Signalling lamps need a code.
- **Lasers** produce a very intense beam of light and can be used for:
  - surgery and dental treatment
  - cutting materials in industry
  - weapon guidance
  - laser light shows
  - communication along **optical fibres**.
- Laser light is a very narrow beam of one colour.

G–E

D–C

Morse code

- The **Morse code** uses a series of dots and dashes to represent *letters* of the alphabet.
  - This code is used by signalling lamps as a series of short and long *flashes of light*.
  - It is an example of a **digital signal**.

D–C

Trapped light

- Light can be **reflected** inside a material such as glass, water and Perspex.
  - This process is called **total internal reflection**.
- Light, laser beams and **infrared** radiation can travel along optical fibres.

Remember!
Optical fibres are solid, not hollow!

Total internal reflection in an optical fibre

G–E

Critical angle

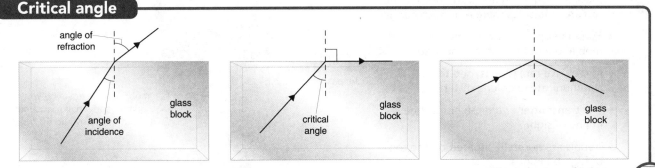

The behaviour of light in an optically dense material

D–C

- When light travels from one material to another, it is normally **refracted**.
- If it is passing from a more **dense** material into a less dense material, the angle of refraction is larger than the angle of incidence.
- When the angle of refraction is 90°, the angle of incidence is called the **critical angle**.
- If the angle of incidence is bigger than the critical angle, the light is reflected:
  - this is total internal reflection.
- Telephone conversations and computer data are transmitted long distances along optical fibres at the **speed** of light (200 000 km/s in glass).
- Some fibres are coated to improve reflection.

Improve your grade

Uses of lasers

Lasers can be used to cut through sheets of metal. A normal white light source will not do this. Explain why. *AO1* [2 marks]

Cooking and communicating using waves

Infrared and microwave radiation

G–E

- All warm and hot bodies emit **infrared radiation**.
  - Hotter bodies emit more radiation than cooler bodies.
  - Dark, dull surfaces absorb more radiation than light shiny surfaces.
  - Dark, dull surfaces emit more radiation than light shiny surfaces.
- **Microwaves** and infrared radiation are part of the **electromagnetic spectrum**.
- Water and fat molecules in food absorb microwaves and heat up.
  - This is the basis of the microwave oven.

Remember!
Black surfaces absorb radiation, they do not attract heat.

Cooking with waves

D–C

- Infrared radiation does not penetrate food very easily.
- Microwaves penetrate up to 1 cm into food.
- Microwaves can penetrate glass or plastic but are **reflected** by shiny metal surfaces:
  - special glass in a microwave oven door reflects microwaves
  - they can cause body tissue to burn.

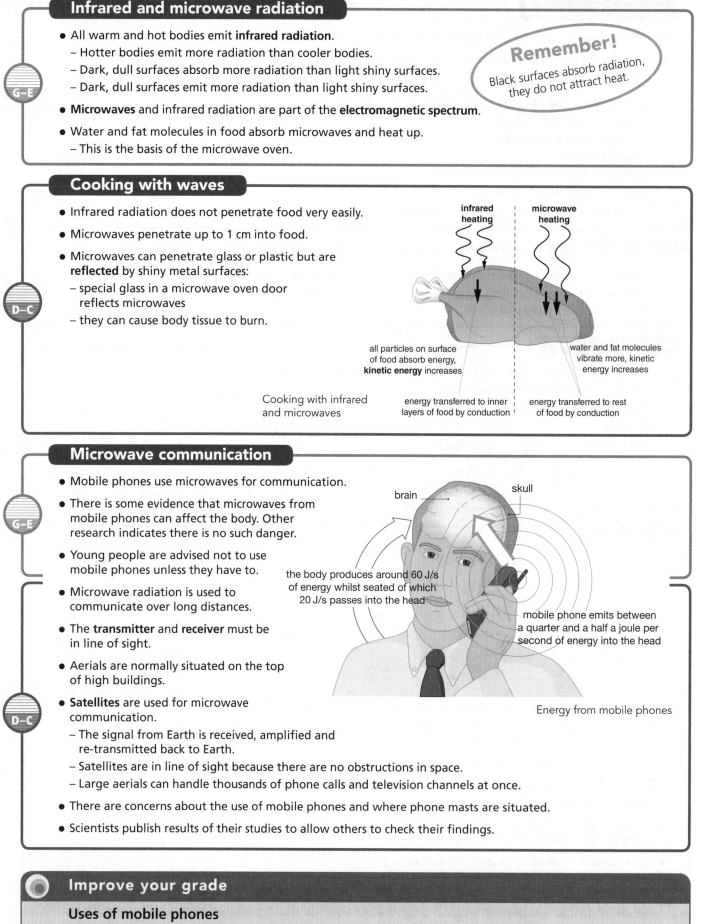

infrared heating   microwave heating

all particles on surface of food absorb energy, **kinetic energy** increases

water and fat molecules vibrate more, kinetic energy increases

energy transferred to inner layers of food by conduction

energy transferred to rest of food by conduction

Cooking with infrared and microwaves

Microwave communication

G–E

- Mobile phones use microwaves for communication.
- There is some evidence that microwaves from mobile phones can affect the body. Other research indicates there is no such danger.
- Young people are advised not to use mobile phones unless they have to.

D–C

- Microwave radiation is used to communicate over long distances.
- The **transmitter** and **receiver** must be in line of sight.
- Aerials are normally situated on the top of high buildings.
- **Satellites** are used for microwave communication.
  - The signal from Earth is received, amplified and re-transmitted back to Earth.
  - Satellites are in line of sight because there are no obstructions in space.
  - Large aerials can handle thousands of phone calls and television channels at once.
- There are concerns about the use of mobile phones and where phone masts are situated.
- Scientists publish results of their studies to allow others to check their findings.

brain   skull

the body produces around 60 J/s of energy whilst seated of which 20 J/s passes into the head

mobile phone emits between a quarter and a half a joule per second of energy into the head

Energy from mobile phones

Improve your grade

Uses of mobile phones
The distance between mobile phone masts can vary between 0.8 km in the centre of a town up to 4 km in the countryside. Why are mobile phone masts much closer together in the centre of towns? *AO1* [1 mark]

Data transmission

Invisible waves

- **Infrared** signals can be used for remote controllers:
  - television and other audio-visual equipment
  - cordless mouse
  - household equipment e.g. garage doors, blinds.
- Passive infrared **sensors** detect body heat and are used in burglar alarms.
- **Digital signals** have two values – on and off.

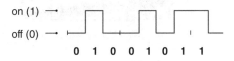

A digital signal has two values

- **Analogue** signals can have any value and are continuously variable.
- The analogue signal changes both its **amplitude** and **wavelength**.

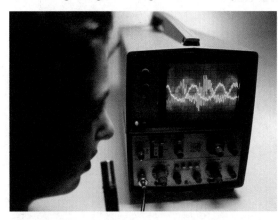

G–E

Digital signals

- The switchover from analogue to digital started in 2009 and is planned to finish by 2015. This may be delayed until more people buy digital radios. The switchover for both radio and TV means:
  - improved signal quality for both picture and sound
  - a greater choice of programmes
  - being able to interact with the programme
  - information services such as programme guides and subtitles.
- Infrared signals carry information that allows electronic and electrical devices to be controlled.

D–C

Optical fibres

- When light passes from a more **dense** material such as glass into a less dense material such as air, the angle of **refraction** is greater than the angle of incidence. If the angle of incidence is too great, the light is **reflected** back into the glass.
- Light, infrared radiation or a **laser** beam can travel along an optical fibre:
  - **Optical fibres** are very flexible.
  - Signals are coded and transmitted digitally.
  - When the light meets the boundary with air, it is reflected back into the fibre.

D–C

Remember!
An optical fibre is solid, not a hollow tube.

Improve your grade

Analogue and digital signals
Describe the differences between analogue and digital signals. *AO1* [2 marks]

Wireless signals

Wireless technology

- Wireless technology means that computers and phones can be used anywhere.
- **Radio waves** are part of the **electromagnetic spectrum**.
- Radio waves can be **reflected**.
  - This means that a signal can be received even if the aerial is not in line of sight.
  - Sometimes an aerial receives a reflected signal as well as a direct signal.
  - On a television picture this appears as ghosting.

Causes of ghosting

Radio refraction and interference

- Radio waves are reflected and **refracted** in the Earth's **atmosphere**:

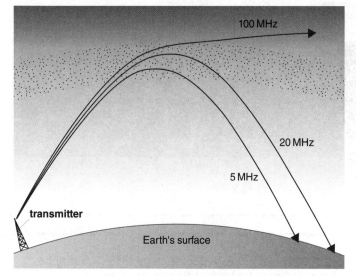

Refraction of waves in the atmosphere

  - The amount of refraction depends on the **frequency** of the wave.
  - There is less refraction at higher frequencies.
- Radio stations broadcast signals with a particular frequency.
- The same frequency can be used by more than one radio station:
  - the radio stations are too far away from each other to interfere
  - but in unusual weather conditions, the radio waves can travel further and the broadcasts interfere.
- **Interference** is reduced if **digital** signals are used.
- Digital Audio Broadcasting or DAB also provides a greater choice of radio stations but the audio quality is not as good as the FM signals currently used.

EXAM TIP

Look at the allocation of marks. There are two marks available here so two different ideas are needed for an answer.

Improve your grade

Radio communication

The picture shows a transmitter and receiver on the Earth's surface, out of line of sight.

Describe how long-wave radio signals travel from the transmitter to the receiver. *AO1* [2 marks]

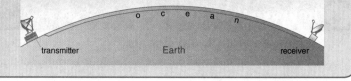

P1 Energy for the home

Stable Earth

Earthquakes

- Earthquakes happen at a *fault*.
- Shock **waves** travel through and round the surface of the Earth, causing damage to buildings and the Earth's surface.
  - Sometimes a **tsunami** is formed.
- We detect earthquakes using a *seismometer*:
  - a heavy weight, with a pen attached, is suspended above a rotating drum with graph paper on it
  - the base is bolted to solid rock
  - during an earthquake, the base moves
  - the pen draws a trace on the graph paper.

G–E

Remember!
It is the base of the seismometer that moves – the weight and pen stay still

Earthquake waves

- The focus is where the earthquake happens below the surface: the *epicentre* is the point on the surface above the focus.
- L waves travel round the surface very slowly.
- **P waves** are **longitudinal** pressure **waves**:
  - P waves travel through the Earth at between 5 km/s and 8 km/s
  - P waves can pass through solids and liquids.
- **S waves** are **transverse** waves:
  - S waves travel through the Earth at between 3 km/s and 5.5 km/s
  - S waves can only pass through solids.
- A seismograph shows the different types of earthquake wave.

D–C

Seismograph trace

Safe sunbathing

- **Ultraviolet radiation** causes a suntan.
- Too much exposure to ultraviolet radiation can cause sunburn, skin **cancer**, cataracts or skin aging.
- Sun screen filters out ultraviolet radiation.
- **Ozone** is found in the **stratosphere**.
- Ozone helps to filter out ultraviolet radiation.
- **CFC gases** from aerosols and fridges destroy ozone and reduce the thickness of the ozone layer.

G–E

Tan or burn

- A tan is caused by the action of ultraviolet light on the skin.
- Cells in the skin produce **melanin**, a pigment that produces a tan.
- People with darker skin do not tan as easily because ultraviolet radiation is filtered out.
- Use a sun screen with a high SPF or sun protection factor, to reduce risks.

  safe length of time to spend in the sun = published normal burn time × SPF

- People are becoming more aware of the dangers of exposure to ultraviolet radiation, including the use of sun beds.

D–C

Ozone depletion

- At first scientists did not believe there was thinning of the ozone layer – they thought their instruments were faulty but other scientists confirmed the results and increased confidence in the findings.

D–C

Improve your grade

Ozone depletion

The ozone layer in the atmosphere is getting thinner. This means more ultraviolet radiation reaches the Earth.

(a) Write down *two* effects on humans of increased exposure to ultraviolet radiation. *AO1* [2 marks]

(b) How can humans reduce the effects of exposure to ultraviolet radiation? *AO1* [1 mark]

P1 Summary

Energy is transferred when a substance changes temperature.

The amount of energy transferred depends on the mass, temperature change and specific heat capacity.

energy transferred = mass × specific heat capacity × temperature change

Heat and temperature

Energy is transferred from a hotter to a colder body.

Temperature can be represented by a range of colours in a thermogram: hot by white/yellow/red; cold by black/dark blue/purple.

Energy is transferred when a substance changes state.

The amount of energy transferred depends on the mass and the specific latent heat.

energy transferred = mass × specific latent heat

Energy transfer

Air is a good insulator and reduces energy transfer by conduction.

Conduction in a solid is by the transfer of kinetic energy.

Trapped air reduces energy transfer by convection.

Convection currents are caused by density changes.

Energy saving in the home can be achieved by:
- double glazing
- cavity wall insulation
- draught strip
- reflecting foil
- loft insulation
- curtains
- careful design.

Shiny surfaces reflect infrared radiation to reduce energy transfer.

Energy transformations can be represented by Sankey diagrams.

$$\text{efficiency} = \frac{\text{useful energy output (} \times 100\%)}{\text{total energy input}}$$

Waves transfer energy

Warm and hot objects emit infrared radiation.

Infrared radiation is used for cooking.

Microwaves can be used for cooking and for communication when transmitter and receiver are in line of sight.

Laser light is single colour and in phase.

Laser light, visible light and infrared are all used to send signals along optical fibres by total internal reflection.

All waves have amplitude, frequency and wavelength

wave speed = frequency × wavelength

Digital and analogue signals are used for communication.

Morse is a digital code.

Analogue signals are continuously variable, digital are on or off.

Digital signals are clearer.

Radio waves, microwaves, infrared, visible light and ultraviolet are some of the waves in the electromagnetic spectrum.

The energy of the wave increases as the wavelength decreases.

All electromagnetic waves can be reflected, refracted and diffracted.

Radio waves are used for communication.

Wireless technology has many advantages.

The stable Earth

Earthquake waves travel through the Earth.

Longitudinal P waves travel faster than transverse S waves.

Exposure to ultraviolet radiation causes sun burn and skin cancer.

Sun screen and sun block reduces damage caused by ultraviolet radiation.

CFCs are causing the ozone layer to become thinner.

Collecting energy from the Sun

Ionisation

- **Photocells** use light and solar cells use sunlight to produce DC electricity.
- The larger the area of the photocell, the more electricity is produced.
- Photocells can be used in places where mains electricity is not available.
- The advantages of photocells are that:
  - they are robust and do not need much maintenance
  - they need no fuel and do not need long power cables
  - they cause no **pollution** and do not contribute to **global warming**
  - they use a **renewable energy** resource.
- The only disadvantage is that they do not produce electricity when it is dark or too cloudy.

G–E

D–C

Energy from the Sun

- Solar water heaters use **energy** from the Sun to heat water:
  - water passes through small tubes over a black plate
  - the plate is black to absorb as much **radiation** as possible
  - hot water passes to a storage tank by convection.

G–E

Passive solar heating

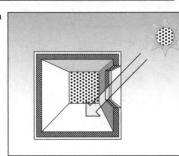 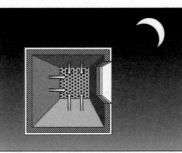

Passive solar heating

> **Remember!**
> In the Southern Hemisphere, the larger windows will need to face North.

G–E

- A house has large windows facing the Sun (South) and small windows facing North:
  - during the day energy from the Sun warms the walls and floors
  - during the night the walls and floors radiate energy back into the room.
- Curved solar reflectors focus energy from the Sun and can be used to directly heat water.

Energy from the wind

- *Convection currents* are formed in the air when there are differences in land **temperature** – this is known as wind.
- Moving air has kinetic energy which is transferred into electricity by a wind **turbine**.
- Wind is a renewable form of energy, but it does depend on the **speed** of the wind:
  - wind turbines do not work if there is no wind, nor if the wind speed is too great.
- Wind farms do not contribute to global warming, nor do they **pollute** the **atmosphere** but they can be noisy, take up a lot of space and people sometimes complain that they spoil the view.

G–E

D–C

EXAM TIP

Always explain your answer when the question requires it.

Improve your grade

Photocells

Describe and explain two situations in which photocells would be the ideal energy source. *AO2* [4 marks]

Generating electricity

Generators

- A **dynamo** is one example of a **generator** in which a magnet rotates inside a coil of wire to produce **alternating current** (AC).
- If a wire is moved near a magnet, or a magnet moved near a wire, a **current** is produced in the wire.

> **Remember!**
> A battery produces direct current (DC).

Bigger currents and voltages

- The current from a dynamo can be increased by:
  - using a stronger magnet
  - increasing the number of turns on the coil
  - rotating the magnet faster.
- The output from a dynamo can be displayed on an **oscilloscope**.
- An oscilloscope trace shows how the current produced by the dynamo varies with time.
- The time for one complete cycle is called the period of the alternating current.

the height of the wave is the maximum (peak) voltage

the length of the wave represents the time for one cycle

Oscilloscope trace

> **Remember!**
> Frequency = 1 ÷ period
> **Frequency** is measured in **hertz** (Hz)
> Period is measured in seconds (s)

Practical generators

- A simple generator consists of a coil of wire rotating between the poles of a magnet:
  - the coil cuts through the magnetic field as it spins
  - a current is produced in the coil.
- A current can be produced if the coil remains stationary and the magnets move.
- Generators at **power stations** work on the same principle.

> **Remember!**
> It is the *relative* movement of magnet and coil that is important.

Power stations

- Most power stations use **fossil fuels** as the **energy** source. Fossil fuels are **non-renewable energy** resources
- The **National Grid** distributes electricity to consumers and energy is lost from the overhead power lines to the surroundings as heat.

- In conventional power stations, fuels are used to heat water:
  - water boils to produce steam
  - steam at high pressure turns a **turbine**
  - the turbine drives a generator.

Energy efficiency

- Energy in a power station is lost in the boilers, generator and cooling towers.
- **Efficiency** is a measure of how well a device transfers energy.

What is the efficiency of a power station if 60 MJ of fuel energy is converted into 20 MJ of electrical energy?

$$\text{efficiency} = \frac{\text{useful energy output}}{\text{total energy input}}$$

$$= \frac{20\,000\,000}{60\,000\,000}$$

$$= 0.33 \text{ or } 33\%$$

Improve your grade

Generators
Describe how electricity is generated. *AO1* [3 marks]

P2 Living for the future (energy resources)

Global warming

Greenhouse gases

- **Carbon dioxide**, water vapour and methane are gases found in the Earth's **atmosphere**. They are examples of **greenhouse gases** which stop heat from the Earth radiating into space. This is known as the greenhouse effect.

- Most **wavelengths** of electromagnetic radiation can pass through the Earth's atmosphere, but **infrared radiation** is absorbed.

- Carbon dioxide occurs naturally in the atmosphere as a result of:
  - natural forest fires
  - volcanic eruptions
  - **decay** of dead plant and animal matter
  - evaporation from the oceans
  - **respiration**.

- Man-made carbon dioxide is caused by burning **fossil fuels**, waste incineration, **deforestation** and **cement** manufacture.

- Water vapour is the most significant greenhouse gas:
  - almost all of the water vapour occurs naturally
  - a mere 0.001% comes from human activity
  - half of the greenhouse effect is due to water vapour and a further quarter is due to clouds.

- Methane is produced when organic matter decomposes in an environment lacking oxygen:
  - natural sources include wetlands, termites, and oceans
  - man-made sources include the mining and burning of fossil fuels, the digestive processes in animals such as cattle, rice paddies and the burying of waste in landfills.

G–E

D–C

The Greenhouse effect

- **Global warming** is the increase in temperature as a result of human activity increasing the levels of greenhouse gases.

- The increase in greenhouse gases is due to:
  - increased **energy** use
  - increased carbon dioxide and methane emissions
  - deforestation.

Rise in CO_2 emissions

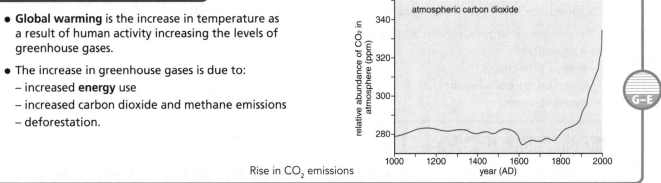

G–E

Dust warms, dust cools

- Dust in the atmosphere can have opposite effects:
  - the smoke from the factories reflects radiation from the town back to Earth. The **temperature** rises as a result
  - the ash cloud from a volcano reflects radiation from the Sun back into space. The temperature falls as a result.

D–C

Scientific data

- Scientists will need to monitor concentrations of greenhouse gases around the world and share their findings. This is the only way we will be able to obtain a full picture of what is happening.

G–E

- The vast majority of scientists agree that the evidence supports climate change. The average temperature of the Earth has increased steadily during the past 200 years.

- What scientists do not agree on is the extent to which human activity has contributed.

- It is important that decisions about what to do about global warming are made on the basis of scientific evidence, not on the basis of unsubstantiated opinions.

D–C

Improve your grade

Earth's temperature
What effects can dust in the atmosphere have on the temperature of the Earth?
AO1 [2 marks]

Fuels for power

Measuring power

- **Power** is a measure of the rate at which **energy** is used.

  power = **voltage** × **current**

> What is the power rating of a kettle working at 9 amps (A) on the mains supply of 230 **volts** (V)?
> power = 9 × 230 = 2070 W

- The unit of power is the **watt** (W) or **kilowatt** (kW).
- The cost of using an appliance depends on its power rating and the length of time it is switched on.
- The unit of electrical energy used in the home is the **kilowatt hour** (kWh).

  cost of electricity used = energy used × cost per kWh

> Calculate the cost of using a 250 W television for 30 minutes if one kWh of electricity costs 10p?
> energy used = power × time = 0.25 × 0.5 = 0.125 kWh
> cost of electricity = energy used × cost per kWh = 0.125 × 10 = 1.25p

Energy sources

- **Fossil fuels** (coal, oil or gas) are the most common energy source for **power stations**.
- Some power stations use renewable **biomass** fuel such as wood, straw or manure.
- **Nuclear power stations** use uranium and sometimes **plutonium** as a fuel.
- Some energy sources are more appropriate than others in a particular situation.

- The choice of energy sources depends on several factors:
  - availability
  - ease of extraction
  - effect on the environment
  - associated risks.

The National Grid

- The **National Grid** is a series of **transformers** and power lines that transport electricity from the power station to the consumer.
- Transformers reduce or increase the voltage of an alternating current.

- In the National Grid, transformers are used to step up the voltage to as high as 40 000 V. The high voltage leads to:
  - reduced energy loss
  - reduced distribution costs
  - cheaper electricity for consumers.
- Transformers are then used again to step down the voltage to a more suitable level for the consumer.

EXAM TIP

When doing calculations always show what equation you are using and the working as you may gain some credit for this.

Improve your grade

Cost of electricity

Calculate the cost of using a 2000 W hairdryer for 30 minutes to dry your hair if one kWh costs 10p. *AO2* [3 marks]

Nuclear radiations

Ionisation

- The three main types of ionising **radiations** are alpha (α), beta (β) and gamma (γ).
- Radiation from nuclear waste causes **ionisation**.
- **DNA** is a chemical in cells which mutates on exposure to radiation and divides uncontrollably. This can cause **cancer**.

- **Atoms** contain the same number of **protons** and electrons – this means they are neutral.
- Ionisation involves gaining or losing **electrons**:
  - when the atom gains electrons, it becomes negatively charged
  - when the atom loses electrons, it becomes positively charged.

G–E

D–C

Handling radioactive material safely

- Keep a safe distance from sources and do not handle directly, always use tongs.
- Use shielding to absorb radiation.
- Use for the minimum time necessary.
- Wear suitable protective clothing when necessary.

G–E

Properties of ionising radiations

- Alpha, beta and gamma radiations come from the **nucleus** of an atom.
- Alpha radiation causes most ionisation and gamma radiation the least.
- Alpha radiation is short-ranged (a few cm) and is easily absorbed by a sheet of paper or card.
- Beta radiation has a range of about 1 m and is absorbed by a few mm of aluminium.

Penetration of three types of radiation — paper — 3 mm thick aluminium — 3 m thick lead block

- Gamma radiation is much more penetrating and, although a few centimetres of **lead** will stop most of the radiation, some can pass through several metres of lead or **concrete**.

D–C

Uses of radioactivity

- Ionising radiation kills cells and living organisms so can be used in cancer treatment called **radiotherapy**.
- Smoke alarms contain a source of alpha radiation:
  - the radiation ionises the oxygen and nitrogen atoms in air which causes a very small electric **current** that is detected. When smoke fills the detector in the alarm during a fire, the air is not so ionised, the current is less and the alarm sounds.
- Thicknesses in a paper rolling mill can be controlled using a source of beta radiation and detector:
  - the amount of radiation passing through the sheet is monitored and the pressure on the rollers adjusted accordingly.
- Gamma radiation kills **microbes** and **bacteria** so can be used for sterilising medical instruments. It can also be used to check for leaks in pipes and welds.
- The passage of blood and other substances can be traced around the body using a beta or gamma source.

G–E

D–C

Nuclear waste

- **Nuclear power stations** do not produce carbon dioxide so do not contribute to **global warming** but the waste is radioactive and harmful.
- **Plutonium** is a waste product from nuclear reactors which can be used to make weapons.
- Some low level **radioactive waste** can be buried in landfill sites. High level waste is encased in glass and buried deep underground or reprocessed.

G–E

D–C

EXAM TIP

Read the question carefully to ensure you have fully understood what is required.

⊙ Improve your grade

Ionising radiation

Describe two *medical* uses of nuclear radiation. AO1 [2 marks]

Exploring our Solar System

Our Solar System

- Earth is just one of the planets that orbit the Sun in our **Solar System**.

- It is easier to see objects in our Solar System in the night sky than it is during the day:
  - we can see **stars** that are far away as they are very hot and produce their own light
  - we see the Moon and sometimes other planets when they reflect light from the Sun.

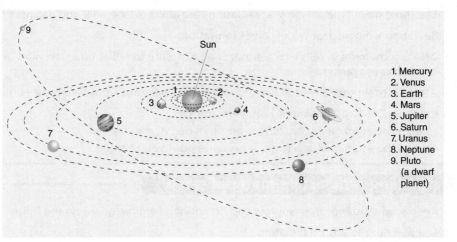

Sun

1. Mercury
2. Venus
3. Earth
4. Mars
5. Jupiter
6. Saturn
7. Uranus
8. Neptune
9. Pluto (a dwarf planet)

The Solar System

- Some stars are in our *galaxy*, which is called the *Milky Way*.

- The **Universe** consists of stars, planets, **comets, meteors, black holes** and many other bodies. Large groups of stars are called galaxies.

- Comets have very **elliptical orbits**. They pass inside the orbit of Mercury and go out well beyond the orbit of Pluto.

- A meteor is made from grains of dust that burn up as they pass through the Earth's **atmosphere**:
  - they heat the air around them which glows and the streak is known as a 'shooting star'.

- Black holes are formed where large stars used to be:
  - you cannot see a black hole because light cannot escape from it; it has a very large **mass** but a very small size.

Exploring the planets

- Unmanned spacecraft (**probes**) have explored the surface of the Moon and Mars to measure **temperature**, gravitational force, **radiation**, magnetic fields and atmosphere.

- Remote vehicles have driven over the surface of Mars. These robots take photographs, and analyse rocks and the atmosphere. One day they will collect rocks to bring back to the Earth.

- The Hubble Space telescope orbits Earth collecting information from the furthest galaxies.

- The Moon is the only body in space visited by humans.

- Unmanned probes can go where conditions are deadly for humans.

- Spacecraft carrying humans need to carry large amounts of food, water and oxygen aboard.

- Astronauts can wear normal clothing in a pressurised spacecraft.

- Outside the spacecraft they need to wear special spacesuits:
  - a dark visor stops an astronaut being blinded
  - the suit is pressurised and has an oxygen supply for breathing
  - the surface of the suit facing towards the Sun can reach 120 °C, whilst that facing away from the Sun may be as cold as −160 °C.

- When travelling in space, astronauts are subjected to lower gravitational forces than on Earth.

A long way to go!

- Distances in space are very large and radio signals take a long time to travel through the Solar System.

- Light travels at 300 000 km/s:
  - one **light-year** is the distance light travels in one year
  - light from the Sun takes about eight minutes to reach us on Earth.

Remember!
A light-year is the distance light travels in one year.

Improve your grade

Travelling into space
Describe the problems astronauts will experience carrying out daily routines when they are aboard the International Space Station. *AO2* [3 marks]

Threats to Earth

Asteroid damage

- When an **asteroid** hits Earth it leaves a large crater.
- The asteroid that collided with Earth 65 million years ago caused more than just a crater:
  - hot rocks rained down
  - there were widespread fires
  - **tsunamis** flooded large areas
  - clouds of dust spread around the world in the upper atmosphere
  - sunlight could not penetrate and **temperatures** fell
  - 70% of all species, including dinosaurs, became extinct.

G–E

Asteroids

- Asteroids are mini-planets or planetoids orbiting the Sun:
  - most orbit between Mars and Jupiter
  - they are large rocks that were left over from the formation of the **Solar System**.

D–C

Origin of the Moon

- Scientists believe our Moon was a result of the collision between two planets in the same orbit.

G–E

- The iron **core** of the other planet melted and joined with the Earth's core, less **dense** rocks began to orbit and they joined together to form our Moon.

D–C

Evidence for asteroids

- Geologists examine evidence to support the theory that asteroids have collided with Earth:
  - near to a crater thought to have resulted from an asteroid impact, they found quantities of the metal iridium – a metal not normally found in the Earth's **crust** but common in meteorites
  - many fossils are found below the layer of iridium, but few fossils are found above it
  - tsunamis have disturbed the fossil layers, carrying some fossil fragments up to 300 km inland.

D–C

Comets

- **Comets** are lumps of ice filled with dust and rock.
- When a comet orbits close to the Sun, the ice warms up and a glowing tail is thrown out which is made up of the dust and rocks freed from the ice.

G–E

- Compared to the near circular orbits of the planets, the orbit of a comet is very elliptical.
- Most comets pass inside the orbit of Mercury and well beyond the orbit of Pluto.
  - As the comet passes close to the Sun, the ice melts and solar winds blow the dust into the comet's tail which always points away from the Sun.

D–C

Pluto

Mercury

Sun

comet

A comet has an elliptical orbit

> **Remember!**
> The longer we study and plot the path of a NEO, the more accurate will be our prediction about its future movement and risk of collision.

NEOs

- Some comets and asteroids have orbits that pass close to the orbit of Earth:
  - these are known as **near-Earth objects** (NEOs)
  - a small change in their direction could bring them on a collision course with Earth.

G–E

- Scientists are constantly monitoring and plotting the paths of comets and other NEOs.
- Scientists monitor the position of NEOs using telescopes although they are sometimes difficult to see.

D–C

Improve your grade

Comets

How do we see a comet? *AO1* [2 marks]

The Big Bang

The birth of the Universe

G–E

- 15 billion years ago, all of the matter in the **Universe** was in a single point.
- It was very hot – millions upon millions of degrees Celsius.
- There was a sudden expansion – the **Big Bang**.
- The Universe expanded rapidly and cooled down.
- Within a few minutes, **electrons**, **protons**, **neutrons** and the elements hydrogen and **helium** had formed.

The expanding Universe

G–E

- The Universe is expanding.

D–C

- Almost all of the galaxies are moving away from each other.
- The furthest galaxies are moving fastest.

Models of the Universe

G–E

- Ptolemy was one of the first to suggest a model of the Universe. This model puts Earth at the centre of everything.
- In 1514, Copernicus suggested an alternative model in which the planets orbit the Sun. This was supported by observations made by Galileo, 100 years later.

D–C

- Today, most scientists accept the Big Bang theory for the formation of the Universe.
- With the help of the newly invented telescope, Galileo observed four moons orbiting Jupiter.
- This confirmed that not everything orbited the Earth and supported Copernicus' idea that planets orbit the Sun.

The birth of a star

G–E

- New **stars** are being formed all the time.
- They start as a swirling cloud of gas and dust.
- Eventually all stars will die.
- Some stars become **black holes** and light cannot escape from a black hole because of the strong gravitational forces.

The end of a star's life

D–C

- A medium sized star, like the Sun, becomes a red giant:
  - the **core** contracts
  - the outer part cools, changes colour from yellow to red and expands
  - gas shells, called planetary nebula, are thrown out
  - the core becomes a white dwarf shining very brightly but eventually cools to become a black dwarf.
- Large stars become red supergiants:
  - the core contracts and the outer part expands
  - suddenly, the core collapses to form a neutron star and there is an **explosion** called a supernova
  - neutron stars are very **dense**
  - remnants from a supernova can merge to form a new star
  - the core of the neutron star continues to collapse, becomes even more dense and could form a black hole.

Improve your grade

Life of a star

List the last three stages a star like our Sun goes through when it is dying.
AO1 [3 marks]

P2 Living for the future (energy resources)

P2 Summary

Kinetic energy from moving air turns the blades on a wind turbine to produce electricity.

Passive solar heating uses glass to help keep buildings warm.

Energy from the Sun

Some gases in the Earth's atmosphere trap heat from the Sun increasing global warming.
Scientists disagree about how much effect humans have on the increase in levels of these Greenhouse gases.

Photocells do not need fuel or cables, need little maintenance and cause no pollution.

A dynamo produces electricity when coils of wire rotate inside a magnetic field.
The size of the current depends on the number of turns, the strength of the field and the speed of rotation.

$$power = voltage \times current$$

In power stations fuels release energy as heat.
Water is heated to produce steam.
The steam drives turbines.
Turbines turn generators.
Generators produce electricity.

$$efficiency = \frac{output\ (\times 100\%)}{total\ energy\ input}$$

Electricity generation

$$energy = power \times time$$

Transformers change the size of the voltage and current.
The National Grid transmits electricity around the country at high voltage and low current. This reduces energy loss.

Nuclear fuels are radioactive.
The radiation produced can cause cancer.
Waste products remain radioactive for a long time.

The main forms of ionising radiation are alpha, beta and gamma.
Their uses depend on their penetrative and ionisation properties.

Medium-sized stars, such as our Sun, were formed from nebulae and will eventually become red giants, white dwarfs and finally black dwarfs.

Planets, asteroids and comets orbit the Sun in our Solar System.

Our Solar System

When two planets collide, a new planet and a moon may be formed.

The Universe consists of many galaxies.
Models of the Universe have changed over time and sometimes these changes take a long time to be accepted.

The Universe is explored by telescopes on Earth and in space.
Large distances mean that it takes a long time for information to be received and makes inter-galactic travel unlikely.

The Universe

Most asteroids are between Mars and Jupiter but some pass closer to Earth.
They are constantly being monitored.
An asteroid strike could cause climate change and species extinction.

Scientists believe that the Universe started with a Big Bang.

Molecules of life

Cell structure

- When cells are looked at under high magnification, it is possible to see small structures in the cytoplasm and in the **nucleus**:
  - In the cytoplasm there are structures called **mitochondria**. This is where **respiration** occurs, providing the **energy** for life processes.
  - In the nucleus there are **chromosomes,** which contain the genetic code.

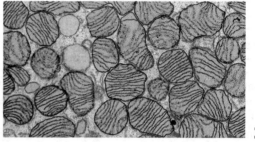

Mitochondria

- The number of mitochondria in the cytoplasm of a cell depends on the activity of the cell. This is because respiration occurs in mitochondria. Cells such as liver or muscle cells have large numbers of mitochondria. This is because the liver carries out many functions and muscle cells need to contract. Both types of cell therefore need lots of energy.

DNA and the genetic code

- The chromosomes in the nucleus are made of **DNA**. This contains the genetic code:
  - The information is in the form of **genes**.
  - Genes control the activity of the cell and the organism by coding for proteins.
- The proteins that a cell produces are needed for growth and repair.

DNA molecule is a double helix; it is like a twisted ladder

pairs of chemicals called **bases** hold the two strands of the DNA molecule together by forming cross-links

- Each gene:
  - is a section of a chromosome made of DNA
  - codes for a particular protein.
- DNA is made of two strands coiled to form a double helix, each strand containing chemicals called **bases**. There are four different types of bases, with cross links between the strands formed by pairs of bases. Each gene contains a different sequence of bases.

The structure of DNA

- Proteins are made in the cytoplasm but DNA cannot leave the nucleus. This means that a copy of the gene needs to be made that can leave the nucleus and carry the code to the cytoplasm.

Discovering the structure of DNA

- The structure of DNA was first worked out by two scientists called James Watson and Francis Crick.

- Watson and Crick built a model of DNA using data from other scientists. Two of the important pieces of data they used were:
  - photographs taken using **x-rays**, which showed that DNA had two chains wound in a helix
  - data indicating that the bases occurred in pairs.

How science works

You should be able to:

- describe examples of how scientists use a scientific idea to explain experimental observations or results
- describe examples of how scientists made a series of observations in order to develop new scientific explanations.

Improve your grade

Genes

Before Watson and Crick's work on DNA, many people used the word 'gene' but did not know what it actually was. Write about what a gene is and what it does.
AO1 [4 marks]

Proteins and mutations

Types of proteins
G–E

- Different cells produce different proteins. Examples of proteins are **collagen**, **insulin** and **haemoglobin**.

- All proteins are made of long chains of **amino acids** joined together.

- Proteins have different functions. Some examples are:
  - structural proteins used to build cells and tissues, e.g. collagen
  - **hormones**, which carry messages to control a reaction, e.g. insulin controls **blood sugar levels**
  - carrier proteins, e.g. haemoglobin, which carries oxygen
  - **enzymes**.

Remember!
Not all hormones are proteins. Insulin is a protein, but others (e.g. testosterone) are not.

D–C

Enzymes
G–E

- Enzymes are protein **molecules** that speed up a chemical reaction. Each enzyme works best at a particular temperature.

- All enzymes have a special area somewhere on the molecule called an **active site.** The chemical that is going to react is called the **substrate** and these molecules fit into the active site when a reaction takes place.

- Enzymes speed up reactions in the body and so are called **biological catalysts**.

- They catalyse chemical reactions occurring in **respiration**, **photosynthesis** and protein synthesis of living cells.

- The substrate molecule fits into the active site of the enzyme like a key fitting into a lock:

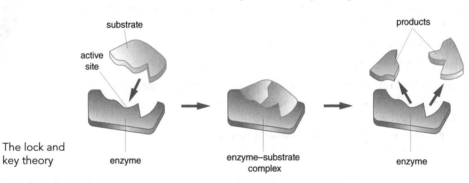

substrate

active site

The lock and key theory

enzyme

enzyme–substrate complex

products

enzyme

D–C

  - This is why enzymes are described as working according to the 'lock and key mechanism'.
  - It also explains why each enzyme can only work on a particular substrate. This is called specificity and it happens because the substrate has to be the right shape.

- Enzymes all work best at a particular **temperature** and **pH**. This is called the optimum. Any change away from the optimum will slow down the reaction.

Mutations
G–E

- Sometimes the genetic material will change. A change to a **gene** is called a gene **mutation**.

- Mutations may occur spontaneously but can be made to occur more often by **radiation** or chemicals.

- When they occur, mutations:
  - may lead to the production of different proteins
  - are often harmful but may have no effect
  - occasionally they might give the individual an advantage.

D–C

Improve your grade

Enzymes
Biological washing powders contain enzymes such as lipase, which break down fat molecules in stains. Explain how enzymes such as lipase work and why biological washing powders should be used at the stated temperature.
AO1 [3 marks], *AO2* [2 marks]

Respiration

Why is respiration important?

- **Respiration** releases **energy** from food. This energy is needed for many processes in living organisms. Examples of processes that require energy from respiration include:
  - muscle contraction to allow organisms to move
  - protein synthesis
  - the control of body **temperature** in **mammals**.

Aerobic respiration

- **Aerobic respiration** involves using oxygen to release the energy from food. The word equation for aerobic respiration is:

  glucose + oxygen → carbon dioxide + water

- The symbol equation for aerobic respiration is:

  $C_6H_{12}O_6 + 6O_2 → 6CO_2 + 6H_2O$

EXAM TIP

When you write this particular equation, make sure that all the letters are capitals and that the numbers are the right size and in the right position.

Anaerobic respiration

- During hard exercise, breathing and heart rate both increase. This is because the muscles are working harder and so are respiring faster. They need more oxygen and there is more **carbon dioxide** to be removed.

- During exercise, despite an increase in breathing rate and heart rate, the muscles often do not receive sufficient oxygen. They start to use **anaerobic respiration** in addition to aerobic respiration.

- The word equation for anaerobic respiration is:

  glucose → lactic acid (+ energy)

- Anaerobic respiration has two main disadvantages over aerobic respiration:
  - The lactic acid that is made by anaerobic respiration builds up in muscles, causing pain and fatigue.
  - Anaerobic respiration releases much less energy per glucose **molecule** than aerobic respiration.

An athlete running long distances tries to use only aerobic respiration

Measuring respiration rate

- It is possible to set up different experiments to measure the rate of respiration. Two ways to do this involve:
  - measuring how much oxygen is used up – the faster it is consumed, the faster the respiration rate
  - the rate at which carbon dioxide is made.

- Scientists can use these results to calculate the **respiratory quotient**. This is worked out using the formula:

  $$RQ = \frac{carbon\ dioxide\ produced}{oxygen\ used}$$

Remember!
The RQ for glucose is 1, from the equation:
$$\frac{6CO_2}{6O_2} = 1$$

Improve your grade

Training

Training improves the blood flow to the muscles of marathon runners. This helps them to respire aerobically rather than anaerobically when they run their races.

Explain why an improved blood supply helps them to do this and what advantage it gives them. *AO1* [2 marks], *AO2* [2 marks]

Cell division

Becoming multicellular

- Many simple organisms such as **bacteria** and many protoctista only have one cell. They are unicellular. Amoebae are unicellular organisms.

- More complex organisms have more than one cell and are **multicellular**.

- There are a number of advantages of being multicellular, as humans are. It allows an organism to become larger and more complex. It also allows different cells to take on different jobs. This is called **cell differentiation**.

Mitosis

- Most body cells contain **chromosomes** in matching pairs. When an organism needs to produce new cells for growth the chromosomes have to be copied.

- The type of cell division that makes new cells for growth is also used for:
  - replacing worn-out cells
  - repairing damaged tissue
  - **asexual reproduction**.

- The process that produces new cells for growth is called **mitosis**.

- The cells that are made by mitosis are genetically identical. Before cells divide, **DNA** replication must take place. This is so that each cell produced still has two copies of each chromosome.

- Body cells in **mammals** have two copies of each chromosome, so they are called **diploid** cells.

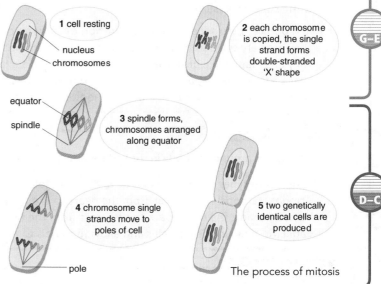

1 cell resting
nucleus
chromosomes

2 each chromosome is copied, the single strand forms double-stranded 'X' shape

equator
spindle
3 spindle forms, chromosomes arranged along equator

4 chromosome single strands move to poles of cell
pole

5 two genetically identical cells are produced

The process of mitosis

Meiosis

- In sexual reproduction **gametes** join together. This is called **fertilisation**.

- Gametes have half the number of chromosomes of body cells. When they join it produces a unique individual because half the **genes** come from each parent.

- Sperm cells are produced in large numbers. This helps to increase the chance of fertilisation because many will not reach the egg.

- The type of cell division that produces gametes is called **meiosis**.

- Gametes are **haploid** cells because they contain only one chromosome from each pair. This means that the zygote gets one copy of a gene from one parent and another copy from the other parent. This produces genetic **variation**.

- The structure of a sperm cell is adapted to its function. It has:
  - many **mitochondria** to provide the energy for swimming to the egg
  - an **acrosome** that releases **enzymes** to digest the egg membrane.

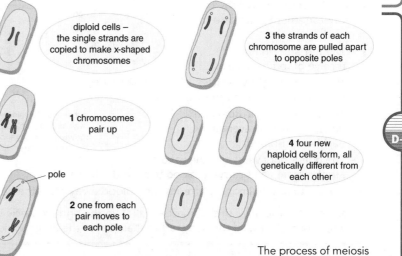

diploid cells – the single strands are copied to make x-shaped chromosomes

3 the strands of each chromosome are pulled apart to opposite poles

1 chromosomes pair up

4 four new haploid cells form, all genetically different from each other

pole
2 one from each pair moves to each pole

The process of meiosis

Improve your grade

Functions of mitosis

Human body cells can only divide by mitosis a certain number of times. Using the functions of mitosis, suggest what effect this has on the body. *AO1* [2 marks], *AO2* [2 marks]

The circulatory system

Blood

G–E

- The blood is made of liquid and several different types of cells. These cells have different functions:
  - **Red blood cells** carry oxygen from the lungs to the tissues.
  - **White blood cells** destroy disease-causing organisms.
  - Platelets help the blood to clot if a blood vessel is damaged.

D–C

- The liquid part of the blood called **plasma** carries a number of important substances around the body:
  - dissolved food substances such as glucose
  - **carbon dioxide** from the tissues to the lungs
  - **hormones** from the glands where they are made, to their target cells
  - plasma proteins such as antibodies
  - waste substances such as urea.

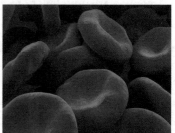

Red blood cells

- Red blood cells are adapted to their function of carrying oxygen in a number of ways:
  - They are very small so that they can pass through the smallest blood vessels.
  - They are shaped like biconcave discs so that they have a large surface area to exchange oxygen quicker.
  - They contain **haemoglobin** to combine with oxygen. It is the haemoglobin that makes them appear red.
  - They don't have a **nucleus** so more haemoglobin can fit in.

Blood vessels

G–E

- Blood moves around the body in three different types of blood vessels called **arteries**, **veins** and **capillaries**.

D–C

- The different types of blood vessels have different jobs:
  - Arteries transport blood away from the heart to the tissues.
  - Veins transport the blood back to the heart from the tissues.
  - Capillaries link arteries to veins and allow materials to pass between the blood and the tissues.

The heart

G–E

- Blood is pumped around the body by the heart. The right side of the heart pumps blood to the lungs and the left side pumps blood to the rest of the body.

- The blood in the arteries is under higher pressure than the blood in the veins and so blood flows from arteries to veins because of this pressure difference.

D–C

- The different parts of the heart work together to circulate the blood:
  - The left and right atria receive blood from veins.
  - The left and right ventricles pump blood out into arteries.
  - The semilunar, tricuspid and bicuspid valves prevent any backflow of blood.
  - The pulmonary veins and the vena cava are the main veins carrying blood back to the heart.
  - The aorta and pulmonary arteries carry blood away from the heart.

RIGHT

pulmonary artery, takes deoxygenated blood to the lungs

aorta, takes oxygenated blood to the body

LEFT

semi-lunar valve

vena cava, brings deoxygenated blood from the body

right atrium

tricuspid valve

valve tendon

pulmonary vein, brings oxygenated blood from the lungs

left atrium

bicuspid valve

right ventricle, thinner wall as pumps blood a relatively short distance to the lungs

left ventricle, has thick muscular wall to pump blood at higher pressure all the way round the body

Blood flow in the heart

- The left ventricle has a thicker muscle wall than the right ventricle because it has to pump blood all around the body rather than just to the lungs, which are close by.

Remember!
The blood in the pulmonary vein is oxygenated and in the pulmonary artery it is deoxygenated, unlike in other veins and arteries.

Improve your grade

Blood vessels and the heart
Explain these observations.

(a) If some blood vessels are cut, the blood oozes out, whereas if others are cut it will pulse out.

(b) The walls of the atria are much thinner than the ventricle walls. *AO2* [4 marks]

Growth and development

Different types of cells

- The different parts of a plant cell perform different jobs:
  - The vacuole contains cell sap and if under pressure it provides support for the plant.
  - The cell wall is made of cellulose, which is strong so it also provides support.
- These structures can be seen in a plant cell by following these steps:
  - Peel a thin layer of cells from an onion.
  - Place the layer on a microscope slide and add a few drops of a stain.
  - Put a cover slip on top and view under a microscope.
- **Bacterial** cells can also be seen using powerful microscopes but they are smaller and simpler than plant and animal cells.

G–E

- Bacterial cells differ from plant and animal cells in that they lack a 'true' **nucleus**, **mitochondria** and chloroplasts.

D–C

Measuring growth

- Growth can be measured as an increase in height, wet mass or dry mass.
- The growth of an individual can be plotted on a growth curve.
- The diagram of a typical growth curve shows the main phases of growth.
- Two of these phases involve rapid growth; one is just after birth and the other in adolescence.
- Dry mass is the best measure of growth.

G–E

D–C

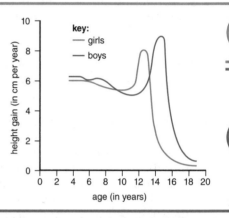

Human growth curve

Differentiation

- Growth involves two main processes. First the cells divide and then they become specialised for different jobs.
- The process of cells becoming specialised is called **cell differentiation**.
- Cells called **stem cells** stay undifferentiated. They can develop into different types of cells.
- Stem cells can be obtained from embryos and could potentially be used to treat some medical conditions.
- However, there are issues arising from stem cell research in animals. Some people think that it is wrong because the embryos are destroyed. Others think that this is acceptable as it can treat life-threatening diseases.

G–E

D–C

> **EXAM TIP**
>
> Make sure that you can give both sides of this argument, even if you have strong views.

Plant and animal growth

- Growth in plants and animals shows different patterns:
  - Animals grow in the early stages of their lives, whereas plants grow continually.
  - All parts of an animal are involved in growth, whereas plants grow at specific parts of the plant.
- Other differences include:
  - Animals tend to only grow to a certain size but many plants can carry on growing.
  - Plant cell division only happens in areas called **meristems**, found at the tips of roots and shoots.
  - The main way that plants gain height is by cells enlarging rather than dividing.
  - Many plant cells keep the ability to differentiate, but most animal cells lose it at an early stage.

G–E

D–C

Improve your grade

Plant growth

Katie wants to investigate the growth of some broad bean seeds.

(a) Describe two different measurements she could use to plot the growth of the young plants.

(b) Describe how the pattern of growth in these plants differs from the growth in her body. *AO1* [4 marks]

New genes for old

Selective breeding

G–E

- Animals and plants can be changed by **selective breeding**. This involves:
  - selecting organisms that best show the desired characteristics
  - cross-breeding these organisms together
  - repeating this selecting and breeding over many generations.

- Selective breeding can help farmers increase the yield of their crops by producing plants that grow quicker, grow larger, and are resistant to diseases and droughts.

D–C

- There are possible problems with selective breeding programmes. They may lead to **inbreeding**, where two closely related individuals mate, and this can cause health problems within the **species**. Certain breeds of dog show these problems.

Genetic engineering

G–E

- **Genes** can be artificially transferred from one living organism to another and this is called **genetic engineering** or genetic modification.

- This transfer of genes can produce organisms with different characteristics.

D–C

- Genetic engineering has advantages and risks:
  - One advantage is that organisms with desired features can be produced very quickly.
  - However, there is a risk that the inserted genes may have unexpected harmful side effects.

- Examples of organisms that have been made using genetic engineering include the following:
  - Rice that contains beta-carotene has been made by inserting the genes that control beta-carotene production from carrots. Humans can then convert the beta-carotene from rice into Vitamin A. This is important because in some parts of the world people rely on rice, which normally has very little Vitamin A.
  - Genetically engineered **bacteria** have been made that produce human **insulin**.
  - Crop plants have been made that are resistant to **herbicides**, frost damage or disease.

- A number of ethical issues are involved in genetic engineering:
  - Some people are worried about possible long-term side effects, e.g. that genetically engineered plants or animals will disturb natural ecosystems.
  - Other people think that it is morally wrong, whatever the intended benefits.

> **Remember!**
> It is the gene that is put into another organism, e.g. the gene for human insulin, not human insulin itself.

Gene therapy

G–E

- In the future it may be possible to use genetic engineering to change a person's genes and cure certain disorders.

D–C

- This process is called **gene therapy**.

Improve your grade

Selective breeding of chickens

Using selective breeding, farmers have produced chickens that lay large eggs.

(a) Describe the steps that would be taken to do this. *AO1* [3 marks]

(b) What might be an advantage of producing these chickens by genetic engineering rather than selective breeding? *AO1* [1 mark]

Cloning

Cloning animals

- Cloning produces genetically identical copies called **clones**:
  - Dolly the sheep was the first **mammal** cloned from an adult.
  - Cloning is an example of **asexual reproduction**.
  - Identical twins are naturally occurring clones.

- Dolly the sheep was produced by a process called **nuclear transfer**. This involves removing the **nucleus** from a body cell and placing it into an egg cell that has had its nucleus removed.

- Animals could be cloned to:
  - mass-produce animals with desirable characteristics
  - produce animals that have been **genetically engineered** to provide human products
  - produce human embryos to supply **stem cells** for **therapy**.

- There are some ethical dilemmas concerning human cloning. Some people think that it is wrong to clone people as they will not be 'true individuals'.

Identical twins are natural clones as they have the same DNA

Remember!

Cloning would not be very successful in recreating endangered or extinct animals, as all the clones would be the same sex and genetically identical.

Cloning plants

- Plants can be grown from cuttings taken from shoots or by **tissue culture**. The plants that are produced are clones.

- A gardener can take a cutting by:
  - cutting a short stem off a plant with a knife
  - dipping the end of the stem into hormone rooting powder
  - putting the cutting into moist sandy soil in a pot
  - covering the plant with a plastic bag to keep the moisture in.

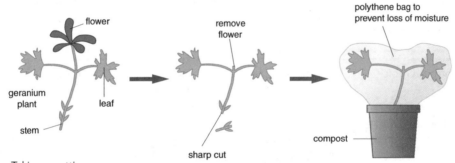

flower

remove flower

polythene bag to prevent loss of moisture

geranium plant

leaf

stem

sharp cut

compost

Taking a cutting

- Spider plants, potatoes and strawberries can all reproduce asexually:
  - Spider plants grow small plants on stems called plantlets. They can grow into a new plant.
  - Strawberries grow long stems called runners that spread over the ground. Small plants bud and grow from the runners at intervals.
  - Potato plants grow a number of tubers underground. Any of these can grow into a new plant.

- Producing cloned plants has advantages and disadvantages:
  - Advantages: growers can be sure of the characteristics of each plant since all the plants will be genetically identical. It is also possible to mass-produce plants that may be difficult to grow from seed.
  - Disadvantages: if the plants become susceptible to disease or to change in the environmental conditions, then all the plants will be affected. There is a lack of genetic **variation** in the plants.

Improve your grade

Taking cuttings

A garden centre wants to sell an attractively coloured geranium plant. They decide to produce many clones of the plant by taking cuttings.

(a) Suggest reasons why they chose this method. AO1 [1 mark]

(b) Give a possible disadvantage of mass-producing plants by this method. AO1 [2 marks]

B3 Summary

Proteins are coded for by DNA by the sequence of bases.

The information in genes is in the form of coded instructions called the genetic code.

Chromosomes are long, coiled molecules of DNA, divided up into regions called genes.

The structure of DNA was first worked out by Watson and Crick.

DNA, proteins and mutations

Enzymes are specific and work by a 'lock and key' mechanism.

Enzyme activity is affected by pH and temperature. This is due to:
- lower collision rates at lower temperatures
- denaturing at extremes of pH and high temperatures.

Proteins:
- are made of long chains of amino acids
- can be structural, hormones, carrier molecules or enzymes.

Mutations may lead to the production of different proteins.

In meiosis, each cell produced has one chromosome from each pair (haploid).

Gametes are produced by meiosis.

New cells for growth are produced by mitosis.

The new cells made by mitosis are genetically identical.

Cell division and growth

There are a number of differences between plant growth and animal growth.

Being multicellular allows organisms to:
- be larger
- use cell differentiation
- be more complex.

Growth can be measured by a change in wet mass, dry mass or length. Dry mass is the best measure.

Undifferentiated cells called stem cells can develop into different cells, tissues and organs.

The symbol equation for aerobic respiration is
$C_6H_{12}O_6 + 6O_2 \rightarrow 6CO_2 + 6H_2O$
The energy released is stored in ATP.

Anaerobic respiration takes place during hard exercise when there is insufficient oxygen available.

Anaerobic respiration produces lactic acid, which builds up in muscles causing pain and fatigue.

Respiration and circulation

Red blood cells carry oxygen around the body.
White blood cells fight infections.
Platelets clot the blood.

The heart has four chambers and is part of a double circulatory system.

Arteries have thick elastic walls and carry blood away from the heart.
Veins carry blood back to the heart.
Capillaries link arteries to veins.

Gene therapy involves changing a person's genes to try and cure disorders.

Dolly the sheep was produced by the process of nuclear transfer – this involves placing the nucleus of a body cell into an egg cell.

Changing genes and cloning

Genetic engineering can be used to produce useful products but raises some ethical issues.

New cloning technology will raise a number of ethical issues.

A selective breeding programme can produce organisms with desired characteristics but may reduce the gene pool leading to problems of inbreeding.

Identical twins are natural clones and some plants reproduce asexually by cloning.

Plants can be cloned by taking cuttings or tissue culture and this provides a number of benefits.

Ecology in the local environment

Ecosystems

- There are:
  - natural ecosystems, such as native woodlands and lakes
  - artificial ecosystems, such as forestry plantations and fish farms.

- An ecosystem, such as a garden, is made up of all the plants and animals living there and their surroundings. Where a plant or animal lives is its **habitat**.

- All the animals and plants living in the garden make up the **community**. The number of a particular plant or animal present in the community is called its **population**.

- Natural ecosystems, such as native woodlands and lakes, have a large variety of plants and animals living there – this means it has good **biodiversity**. Artificial ecosystems, such as forestry plantations and fish farms, have poor biodiversity.

Remember!
Biodiversity is the variety of plants and animals in a habitat.

G–E

D–C

Distribution of organisms

- The distribution of an organism is affected by other organisms, such as predators, and by physical factors such as altitude and **temperature**.

- The distribution of organisms can be mapped using a **transect** line. A long length of string is laid across an area such as a path or seashore. At regular intervals the organisms in a square frame called a quadrat can be counted (for animals) or assessed for percentage cover (for plants). The data can be displayed as a **kite diagram**.

Kite diagram showing distribution of organisms near a path

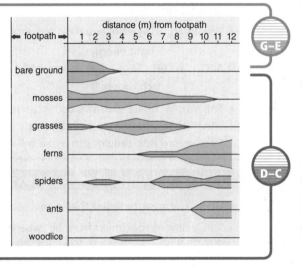

G–E

D–C

Population size

- Various methods are used to collect animals, including:
  - pooters, to suck up small animals
  - nets, to catch flying insects
  - pitfall traps, to catch small crawling insects.

- Quadrats (often 1 m²), can be used to sample plants and animals in a habitat.

- Identification keys are used to identify plants and animals.

- Population size can be estimated by obtaining data from a small sample and scaling up. For example, Shabeena has a large lawn with some dandelions in it. She counts the number of dandelions in ten 1 m² quadrats and calculates the mean as 8 dandelions/m². Her lawn is 100 m². She calculates that there are 800 dandelions in her lawn.

- Shabeena also wants to estimate the woodlouse population in her garden. She uses a capture–recapture method.

$$\text{Population size} = \frac{\text{number in 1st sample} \times \text{number in 2nd sample}}{\text{number in 2nd sample previously marked}}$$

- Shabeena captures 36 woodlice in a pitfall trap, marks them with a white dot and releases them. Later she captures 48 woodlice, 6 with a white dot.

$$\text{Population size} = \frac{36 \times 48}{6} = 288 \text{ woodlice}$$

EXAM TIP

You are expected to know this equation for estimating population size.

D–C

D–C

Improve your grade

Kite diagrams
Look at the kite diagram above.
What do kite diagrams show? Describe trends shown in the kite diagram. *AO1/2* [4 marks]

Photosynthesis

Plants and photosynthesis

- **Photosynthesis** is the process by which a green plant produces glucose.
- The word equation for photosynthesis is:

  light energy
  carbon dioxide + water → glucose + oxygen
  chlorophyll

- Light **energy** and **chlorophyll** (a green pigment) are needed to make the process work.
- Oxygen is a waste gas from photosynthesis.
- As glucose (a simple sugar) is **soluble** in water, it can be carried around the plant. It can be converted into insoluble starch for storage.
- Glucose and starch can be made into other substances and used for growth, storage and energy.
- The **balanced symbol equation** for photosynthesis is:

  light energy
  $$6CO_2 \; + \; 6H_2O \; \rightarrow \; C_6H_{12}O_6 \; + \; 6O_2$$
  chlorophyll

- The simple sugars such as glucose can be:
  - used in **respiration**, releasing energy
  - converted into cellulose to make cell walls
  - converted into proteins for growth and repair
  - converted into starch, fats and oils for storage.

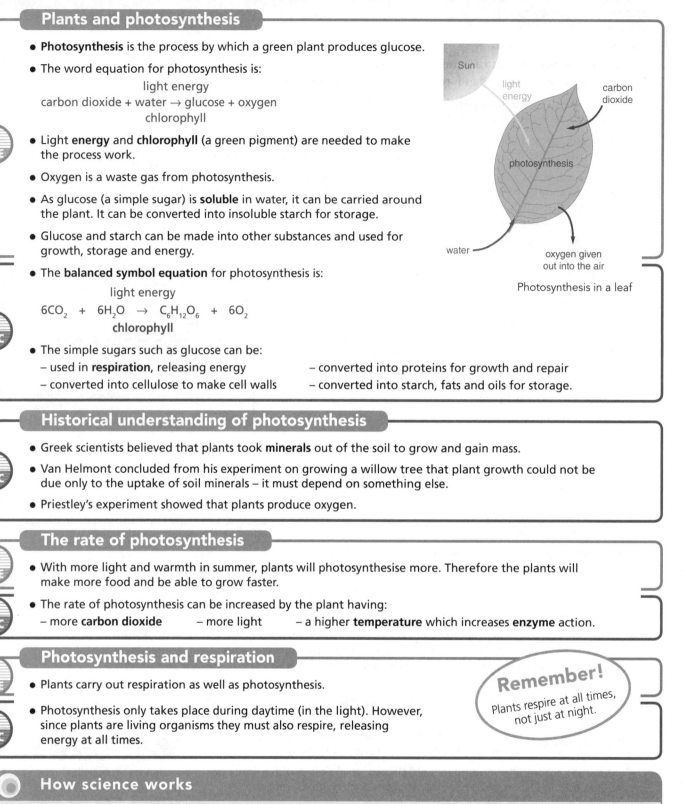

Photosynthesis in a leaf

Historical understanding of photosynthesis

- Greek scientists believed that plants took **minerals** out of the soil to grow and gain mass.
- Van Helmont concluded from his experiment on growing a willow tree that plant growth could not be due only to the uptake of soil minerals – it must depend on something else.
- Priestley's experiment showed that plants produce oxygen.

The rate of photosynthesis

- With more light and warmth in summer, plants will photosynthesise more. Therefore the plants will make more food and be able to grow faster.
- The rate of photosynthesis can be increased by the plant having:
  - more **carbon dioxide**
  - more light
  - a higher **temperature** which increases **enzyme** action.

Photosynthesis and respiration

- Plants carry out respiration as well as photosynthesis.
- Photosynthesis only takes place during daytime (in the light). However, since plants are living organisms they must also respire, releasing energy at all times.

Remember!
Plants respire at all times, not just at night.

How science works

You should be able to:
- explain how a scientific idea, such as an explanation of photosynthesis, has changed as new evidence has been found
- distinguish between claims/opinions and scientific evidence.

Improve your grade

Plant growth in summer
(a) Explain why plants grow better in summer than in winter. *AO1* [2 marks]
(b) Suggest how this information is useful to gardeners. *AO2* [2 marks]

Leaves and photosynthesis

Chloroplasts and photosynthesis

- The green colour of most plants is caused by green chloroplasts inside some cells.
- Some leaf cells are specialised to do other jobs and do not carry out **photosynthesis**.
- The **chlorophyll** pigment in chloroplasts absorbs light **energy** for photosynthesis.
- Photosynthesis also needs:
  - water, which enters through root hairs
  - **carbon dioxide**, which enters through **stomata** (small openings in leaf surfaces).
- Oxygen produced by photosynthesis exits through stomata.

Remember!
Not all leaf cells photosynthesise.

G–E

Leaf adaptations for photosynthesis

- Broad leaves have a large surface area, absorbing a lot of sunlight.

G–E

- A green leaf has many specialised cells, as shown in the diagram.

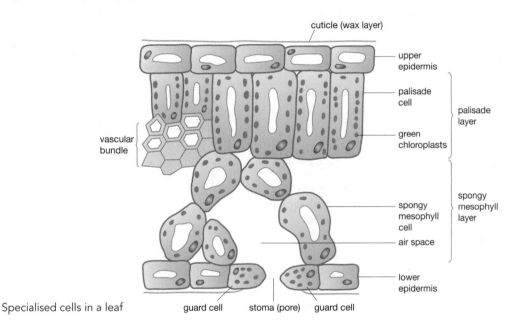

Specialised cells in a leaf

D–C

- Leaves are adapted so that photosynthesis is very efficient:
  - They are usually broad, so that they have a large surface area to get as much light as possible.
  - They are usually thin, so that gases can **diffuse** through easily and light can get to all cells.
  - They contain chlorophyll and other pigments, so that they can use light from a broad range of the light spectrum.
  - They have a network of **vascular bundles (veins)** for support and transport of chemicals such as water and glucose.
  - They have specialised guard cells which control the opening and closing of stomata therefore regulating the flow of carbon dioxide and oxygen as well as water loss.

Improve your grade

Leaf adaptations
Explain why a plant needs chlorophyll and why it needs other pigments.
AO1 [4 marks]

Diffusion and osmosis

Diffusion

G–E
- Substances move in and out of cells by **diffusion** through the **cell membrane**.
- **Carbon dioxide** and oxygen diffuse in and out of plants through their leaves.

D–C
- Diffusion is the net movement of particles in a gas or liquid from an area of high concentration to an area of low concentration, resulting from the **random** movement of the particles.
- This explains how **molecules** of water, oxygen and carbon dioxide can enter and leave cells through the cell membrane. If a plant cell is using up carbon dioxide, there is a lower concentration of it inside the cell, so carbon dioxide will enter by diffusion.
- Leaves are adapted to increase the rate of diffusion of carbon dioxide and oxygen by having:
  - (usually) a large surface area
  - specialised openings called **stomata**, which are spaced out
  - gaps between the **spongy mesophyll cells**.

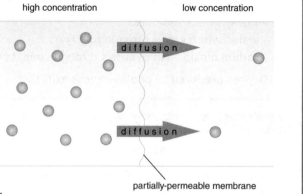

Diffusion of gases into a cell

Osmosis in plants

G–E
- Water can move in and out of plant cells by **osmosis** through the cell membrane.
- The cell walls of plant cells support the cells.
- A plant will droop (wilt) if there is a shortage of water since plant cells will start to collapse.

D–C
- Osmosis is a type of diffusion; it depends on the presence of a **partially-permeable membrane** that allows the passage of water molecules but not large molecules like glucose.
- Osmosis is the movement of water across a partially-permeable membrane from an area of high water concentration (a dilute solution) to an area of low water concentration (a concentrated solution).
- The entry of water into plant cells increases the pressure pushing on the cell wall, which is rigid and not elastic. This **turgor pressure** supports the cell, stopping it, and the whole plant, from collapsing. When too much water leaves a cell, it loses this pressure and the plant wilts.

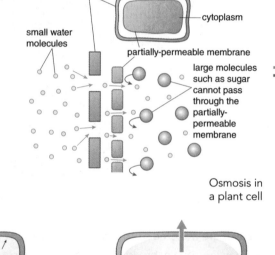

Osmosis in a plant cell

Turgor in plant cells

Plasmolysis in plant cells

Osmosis in animals

G–E
- There are no cell walls around animal cells.
- Water can move into and out of animal cells through the cell membrane.

D–C
- Animal cells also react to intake and loss of water due to osmosis. They will also shrink and collapse when they lose too much water, and swell up when too much water enters.

Remember!
Osmosis is a type of diffusion. It involves the movement of water through a partially permeable membrane.

Improve your grade

Gases for plant growth
(a) Which gases does a green plant need?
(b) How is a leaf adapted to ensure a good supply of these gases? *AO1* [5 marks]

Transport in plants

Water transport

- Water is absorbed by root hairs (extensions of root cells).
- Water is transported from the roots to the leaves.
- Water **evaporates** from the leaves. This is called **transpiration**.
- Healthy plants must balance their water uptake and water loss to avoid wilting.
- **Xylem** and **phloem** are made up of specialised plant cells. Both types of tissue are continuous from the root, through the stem and into the leaf.
- Xylem and phloem cells form **vascular bundles** in dicotyledonous (broad-leaved) plants.
- Xylem cells carry water and **minerals** from the roots to the leaves and are therefore involved in transpiration. Phloem cells carry food substances such as sugars up and down stems to growing and storage tissues. This transport of food substances is called translocation.

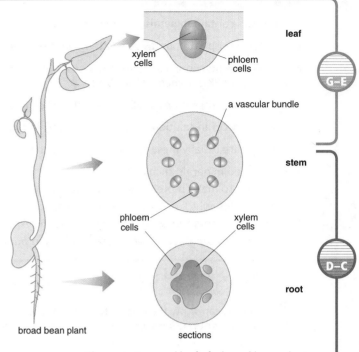

leaf — xylem cells / phloem cells

a vascular bundle

stem

phloem cells / xylem cells

root

broad bean plant

sections

The root, stem and leaf of a broad bean plant

G–E

D–C

Transpiration

- The rate of transpiration is affected by light intensity, **temperature**, air movement (wind) and humidity (the amount of moisture already in the air).

EXAM TIP

Expect a question about how a named factor increases or decreases the transpiration rate.

G–E

- Transpiration is the evaporation (changing from a liquid into a gas) and **diffusion** of water from inside leaves. This loss of water from leaves helps to create a continuous flow of water from the roots to the leaves in xylem cells.
- Root hairs are projections from root hair cells. They produce a large surface area for water uptake by **osmosis**.
- Transpiration ensures that plants have water for cooling by evaporation, **photosynthesis** and support from cells' **turgor pressure**, and for transport of minerals.
- The rate of transpiration is increased by an increase in light intensity, temperature and air movement and a decrease in humidity (the amount of water vapour in the atmosphere).
- The structure of a leaf is adapted to prevent too much water loss, which could cause wilting. Water loss is reduced by having a waxy cuticle covering the outer epidermal cells and by most **stomatal** openings being situated on the shaded lower surface.

Remember!
Water is absorbed by root hairs, not by roots.

D–C

Improve your grade

Transpiration
Explain how water from the soil reaches plant leaves. *AO1* [4 marks]

Plants need minerals

Use of minerals

G–E

- **Fertilisers** contain **minerals** that plants need.
- Nitrates, phosphates, potassium and magnesium compounds are needed for plant growth.
- The relative percentages of nitrates, phosphates and potassium in a fertiliser are shown as NPK values. The letters NPK are chemical symbols.
- The relative proportions of NPK in a fertiliser can be calculated from data.
- Minerals occur in solution (dissolved in water) in the soil. They are taken up by root hairs.
- Minerals are usually present in the soil in low concentrations.

D–C

- Plants need minerals, such as:
  - nitrates, to make proteins, which plants use for cell growth
  - phosphates, which are involved in **respiration** and growth
  - potassium compounds, which are involved in respiration and **photosynthesis**
  - magnesium compounds, which are involved in photosynthesis.

Mineral deficiency

G–E

- Experiments can show the effects that result from plants not getting different minerals. They can be set up using a soil-less culture method, with each **trial** lacking in one mineral.

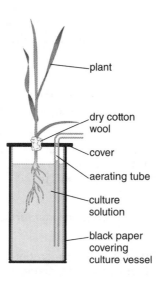

- plant
- dry cotton wool
- cover
- aerating tube
- culture solution
- black paper covering culture vessel

Investigating mineral deficiencies

D–C

- The lack of certain minerals results in specific symptoms:
  - Lack of nitrate causes poor growth and yellow leaves.
  - Lack of phosphate causes poor root growth and discoloured leaves.
  - Lack of potassium causes poor flower and root growth and discoloured leaves.
  - Lack of magnesium causes yellow leaves.

Remember!
The lack of a mineral results in specific changes in plant growth and appearance.

Improve your grade

Mineral deficiency

A fertiliser bag has a label containing NPK information. Explain why this information is important. *AO1* [4 marks]

Decay

The importance of decay

- **Decay** is the breakdown of dead organisms into simpler chemical substances. These act as **fertilisers** for plants and are important for plant growth.

- **Microorganisms** such as fungi and **bacteria** can cause decay. They are called **decomposers**. They need oxygen, water and a warm **temperature**.

- Decomposers are useful because they break down human waste (sewage) and plant waste (**compost**).

- Experiments to show how decomposers cause decay can be carried out.

bungs

thread

this soil loses mass

bag of fresh soil

bag of heated soil

this limewater turns milky

limewater

Experiment to show how decomposers cause decay

- Earthworms, maggots and woodlice are called **detritivores** because they feed on dead and decaying material (**detritus**).

- Detritivores increase the rate of decay by breaking up the detritus and so increasing the surface area for further microbial breakdown.

- The rate of decay can be increased by increasing the temperature, amount of oxygen and water.

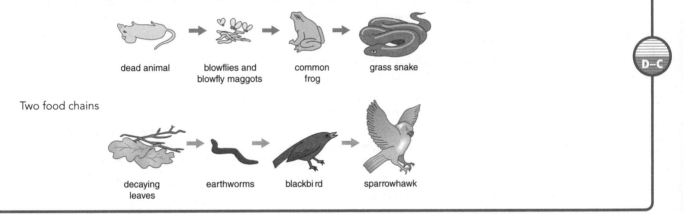

| dead animal | blowflies and blowfly maggots | common frog | grass snake |

Two food chains

| decaying leaves | earthworms | blackbird | sparrowhawk |

Food preservation

- Many different techniques preserve food by reducing the rate of decay. Examples include canning, cooling, freezing, drying, adding salt or sugar, adding vinegar.

- These techniques work in slightly different ways:
  - In canning, foods are heated to kill bacteria and then sealed in a **vacuum** to prevent entry of oxygen and bacteria.
  - Cooling foods will slow down bacterial and fungal growth and reproduction.
  - Freezing foods will kill some bacteria and fungi and slow down growth and reproduction in others.
  - Drying foods removes water so bacteria cannot feed and grow.
  - Adding salt or sugar will kill some bacteria and fungi, as the high osmotic **concentration** will remove water from them.
  - Adding vinegar will produce very acid conditions killing most bacteria and fungi.

Remember!
Keeping food cool in a refrigerator only slows down the rate of decay. It does not stop it completely.

EXAM TIP
Fungi and bacteria do not use **photosynthesis** to make food.

Improve your grade

Decay and decomposers

Describe an experiment to show that decay is caused by decomposers such as bacteria and fungi. *AO1* [4 marks]

Farming

Pest control

G–E

- Pests are any organisms that damage crops.
- Farmers can produce more food when they control pests by using chemical pesticides. However, pesticide use can cause harm to the environment and to health.
- Examples of pesticides include:
  – **insecticides**, which kill insects   – **fungicides**, which kill fungi
  – **herbicides**, which kill unwanted plants (weeds).
- Pests can be also controlled by using other living organisms that are **predators**. This is called **biological control**.

> **Remember!**
> Insecticides, fungicides and herbicides are all pesticides.

D–C

- The use of pesticides such as insecticides, fungicides and herbicides has disadvantages:
  – They can enter and accumulate in food chains causing a lethal dose to predators.
  – They can harm other organisms living nearby which are not pests.
  – Some are persistent (take a very long time to break down and become harmless).

Organic farming

D–C

- Organic farming does not use artificial **fertilisers** or pesticides.
- It uses animal manure and **compost** (instead of artificial fertilisers), **crop rotation** (to avoid build-up of soil pests), nitrogen-fixing crops as part of the rotation, and varying seed planting times to get a longer crop time and avoid certain times of the life cycle of insect pests.
- It avoids expensive fertilisers and pesticides and their disadvantages. However, the crops are smaller and the produce more expensive. Many people believe that organic crops are healthier and tastier than other crops.

Biological control

D–C

- Biological control uses living organisms to control pests. Examples are using ladybirds and certain wasp **species** to eat aphids, which damage plants.
- Biological control can avoid the disadvantages of artificial insecticides and as living organisms are used, once introduced they usually do not need replacing.
- However, many attempts at biological control have caused other problems such as the introduced species eating other useful species. The predator species have then undergone a rapid increase in their **population** so they become pests and then spread into other areas or countries, e.g. the use of cane toads in Australia.
- Introducing a species into a **habitat** to kill another species can affect the food sources of other organisms in a food web, causing unexpected results.

Intensive farming

G–E

- **Intensive farming** methods try to produce as much food as possible from the available land. They improve productivity.
- Examples of intensive farming are:
  – fish farming (keeping large numbers of fish in underwater cages)
  – glasshouses, which can control and give the best conditions for plant growth
  – **hydroponics** (growing plants without using soil)
  – battery farming (keeping large numbers of animals in small spaces).

D–C

- Intensive farming, which makes use of artificial pesticides and fertilisers, is very **efficient** at producing large crop yields cheaply. However, intensive farming methods raise concerns about animal cruelty, as animals are kept in small areas, and about the effects of extensive use of chemicals on soil structure and other organisms.
- Plants can be grown without soil using hydroponics. This system uses a regulated recycling flow of aerated water containing **minerals** and is usually done in glasshouses and polytunnels.
- Hydroponics is a type of intensive farming that is especially useful in areas of barren soil or low rainfall. Tomatoes are a common crop from hydroponics.

Improve your grade

Hydroponics

(a) Describe how plants can be grown without soil. *AO1* [2 marks]
(b) Explain why this method is used in some areas of the world. *AO1* [3 marks]

B4 Summary

An ecosystem, e.g. a garden, includes all living things and their surroundings. Where an organism lives is called its habitat.

Capture–recapture data can be used to calculate a population size estimate.

Population size = $\dfrac{\text{number in 1st sample} \times \text{number in 2nd sample}}{\text{number in 2nd sample previously marked}}$

A community is made up of the organisms living there. A population is the number of a particular organism in a community.

Ecology in the local environment

The distribution of organisms in a habitat is affected by other living organisms and physical factors.

Collecting/counting methods include pooters, nets, pitfall traps and quadrats.

Biodiversity is the variety of different species living in a habitat.

The distribution of organisms can be mapped using a transect line and displayed as a kite diagram

Greek scientists (plants take minerals from the soil), van Helmont (plant growth needs more than minerals) and Priestley (plants produce oxygen) all improved the understanding of photosynthesis.

The word equation for photosynthesis is

carbon dioxide + water $\xrightarrow[\text{chlorophyll}]{\text{light energy}}$ glucose + oxygen

The balanced symbol equation for photosynthesis is

$6CO_2 + 6H_2O \xrightarrow[\text{chlorophyll}]{\text{light energy}} C_6H_{12}O_6 + 6O_2$

Photosynthesis

Leaves are adapted for efficient photosynthesis, (large surface area, thin, contain pigments, have vascular bundles)

Chlorophyll pigments in chloroplasts absorb light energy in photosynthesis.

Broader leaves enable more light energy to be absorbed.

Diffusion is the net movement of particles from a high concentration to a low concentration.

Diffusion and osmosis

Osmosis is the net movement of water from a high to a low water concentration across a partially-permeable membrane.

Carbon dioxide and oxygen diffuse in and out of leaves

Transpiration is the evaporation and diffusion of water from leaves.

Water moves in and out of plant cells by osmosis through the plant membrane.

Plants require nitrates for cell growth, phosphates for respiration and growth, potassium for respiration and magnesium for photosynthesis.

NPK data on fertiliser sacks can be interpreted to plan for improved plant growth.

Experiments can be designed to show the effects on plants of mineral deficiencies.

Minerals and farming

Minerals are dissolved in solution in the soil and are absorbed by root hairs.

Organic farming uses animal manure, crop rotation, weeding and differing planting times.

Biological control has advantages (no chemical pesticides) and disadvantages (introduced predator may become a pest).

Detritivores feed on dead and decaying material.

Rate of reaction (1)

Speed of reaction

- In a chemical reaction **reactants** are made into **products**.   reactants → products
- Some reactions are very fast and others are very slow.
  - Rusting is a very slow reaction.
  - Burning and explosions are very fast reactions.
- The reaction time is the time taken for the reaction to finish. The shorter the reaction time, the faster the reaction.

Reaction rates

- The **rate of reaction** measures how much product is formed in a fixed period of time.
- Reactions are usually fast at the start and then slow down as the reactants are used up.
- The units commonly used when measuring the rate of reaction are:
  - g/s (grams per second) and g/min (grams per minute) when measuring the **mass** of a product formed. g/s is used for faster reactions and g/min for slower reactions.
  - cm^3/s (centimetre cubed per second) and cm^3/min (centimetre cubed per minute) when measuring the volume of gas produced. cm^3/s is used for faster reactions and cm^3/min for slower reactions.

> **EXAM TIP**
>
> Always include units when giving answers to questions involving numbers.

Measuring the rate of reaction

- In reactions that release a gas, the reaction rate can be measured by:
  - monitoring the decrease in mass in a given time period
  - collecting the gas made in a given time period.
- The rate of reaction between magnesium and dilute hydrochloric acid can be easily measured in the laboratory.
- During the reaction:
  - the magnesium ribbon fizzes in dilute hydrochloric acid
  - colourless bubbles of hydrogen are given off.
- The volume of hydrogen collected in the gas syringe can be measured every 10 seconds.

magnesium + hydrochloric acid → magnesium chloride + hydrogen

Measuring the rate of a reaction

- The results of the experiment can be plotted on a graph, as in the example below.
  - The two lines represent two reactions using different quantities of magnesium.
  - The **gradient** (slope) of the lines show how fast each reaction is.
  - The reaction stops because one of the reactants runs out.
- The **limiting reactant** is the reactant *not* in excess that gets used up by the end of the reaction.
- The amount of product formed in a reaction is *directly proportional* to the amount of the limiting reactant used. For example, if the amount of limiting reactant doubles, the amount of product formed doubles (as shown in the graph on the right).

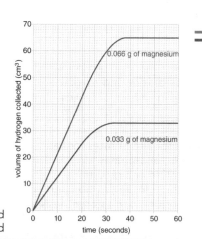

Doubling the limiting reactant used doubles the amount of product formed

Improve your grade

Change in rate of reaction
Limestone is reacted with excess hydrochloric acid in a flask. Explain why the mass of the flask and contents decreases and then the reaction stops. *AO1* [2 marks]

Rate of reaction (2)

Reacting particle model

- Chemical reactions take place when **reactant** particles hit or collide with each other.

- A reaction can be made to go faster by:
  - increasing the **temperature**
  - increasing the **concentration**
  - increasing the *pressure* (if the reactants are gases).

- The **rate of reaction** depends on the number of collisions between the reacting particles:
  - the higher the number of collisions that take place, the faster the reaction.

- The rate of reaction increases with *concentration*, raising the *temperature* or the *pressure* because:
  - with increased concentration, the particles become more crowded. This increases the number of collisions between reacting particles.
  - with increased temperature, the particles gain **kinetic energy**. They move around more quickly, so collisions between reacting particles are more successful, and occur in greater number. (The size of the arrows on the diagram indicate the amount of kinetic energy the particles have.)
  - with increased pressure, the particles are forced closer together, increasing the rate of reaction.

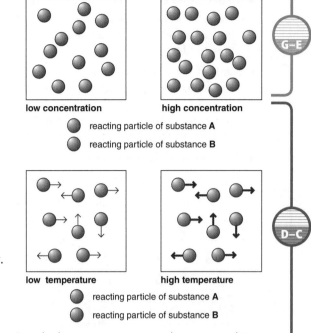

low concentration — high concentration

- reacting particle of substance **A**
- reacting particle of substance **B**

low temperature — high temperature

- reacting particle of substance **A**
- reacting particle of substance **B**

Particle diagrams representing changing conditions

G–E

D–C

Interpreting data on rate of reaction

- The effect of changing conditions on the rate of a reaction can be measured.

- Results can be obtained by collecting gas in a syringe or in a measuring cylinder.
  The graph on the right shows the shape of the line produced by these results.

Rate of gas collected during a reaction

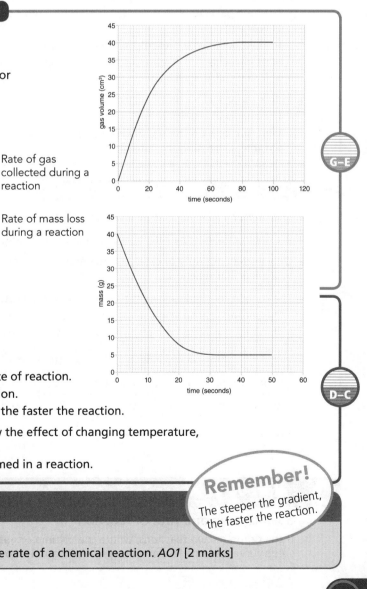

- Results can be obtained by measuring the **mass** loss. The graph on the right shows the shape of the line produced by these results.

Rate of mass loss during a reaction

G–E

- Both tables and graphs can show when a reaction has finished.
  - On a graph the line will be horizontal.
  - In a table the numbers will stop changing.

- Both tables and graphs can be used to compare the rate of reaction.
  - On graphs, the steeper the line, the faster the reaction.
  - In a table, the larger the change between readings, the faster the reaction.

- An essential skill is to be able to sketch graph to show the effect of changing temperature, concentration or pressure on:
  - the rate of reaction and the amount of **product** formed in a reaction.

D–C

Remember!
The steeper the gradient, the faster the reaction.

Improve your grade

Temperature and rate of reaction
Explain the effect of changing the temperature on the rate of a chemical reaction. *AO1* [2 marks]

Rate of reaction (3)

Explosions and surface area

- An **explosion** is a very fast chemical reaction that makes and releases large amounts of gases.
  - The gases move outwards from the reaction at great speed, which causes the explosive effect.
  - Powdered **reactants** always react faster than lumps. Breaking a substance into smaller bits makes the reaction faster.

- Combustible powders can cause explosions.
  - Some powders react with oxygen to make large volumes of **carbon dioxide** and water vapour.
  - The owners of a factory using combustible powders such as flour, custard powder or sulfur must be very careful. They must ensure that the powders cannot reach the open atmosphere and that the chances of a **spark** being produced near the powders is very small.

- Breaking up a block into smaller pieces increases the surface area.

- A powdered reactant has a much larger surface area than the same **mass** of reactant in block form.

- As the surface area of a solid is increased so is the **rate of reaction**.

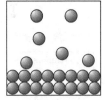 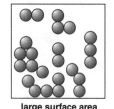

small surface area   large surface area

Changing surface area

- In the 'small surface area' diagram:
  - The blue particles at the bottom represent a solid block of reactant.
  - The red particles represent the other reactant.
  - Most particles in the solid block cannot react as collisions can only occur at the surface.

- In the 'large surface area' diagram:
  - The blue particles from the block are now spread out.
  - More surfaces on these particles are exposed, so more collisions are possible, increasing the rate of reaction.

> **Remember!**
> Surface area increases when a block of solid is cut up into pieces.

Catalysts and rates of reaction

- The rate of a reaction can be increased if a **catalyst** is added.

- A catalyst changes the rate of reaction and is unchanged at the end of the reaction.

- Only a small quantity of catalyst is needed to catalyse a large mass of reactants; each catalyst is specific to a particular reaction.

- The data in the table shows the effect of adding a catalyst to a reaction.
  - The second reaction finishes at 20 seconds, whereas the first reaction finishes at 40 seconds.
  - The rate of reaction is highest in the first 10 seconds of the second reaction.

Time in seconds	10	20	30	40	50
volume of gas with no substance cm³	18	30	60	100	100
volume of gas with catalyst cm³	70	100	100	100	100

Improve your grade

Surface area and rate of reaction

If the graph of two reactions, where the volume of gas is measured over time, shows one steep curve and one less steep curve, which graph shows the quicker reaction? *AO1* [1 mark]

Reacting masses

Relative formula mass

G–E

- The **relative atomic mass** of **elements** can be found in the **periodic table** (see page 248). Atomic mass (A_r) is the larger of the two numbers shown for each element. These add up to give the *relative formula mass* (M_r). For example, the M_r of carbon dioxide, CO_2, is $12 + (2 \times 16) = 44$.

- The relative atomic masses must be added up in the correct order, if there are brackets in the formula, in order to arrive at the correct M_r.

D–C

To find the relative formula mass of calcium nitrate, $Ca(NO_3)_2$ from the relative atomic masses of the elements (A_r) Ca = 40, N = 14, O = 16:

1 Work out the masses inside of the bracket. $14 + (16 \times 3) = 62$
2 Multiply the total by the number outside the bracket. $62 \times 2 = 124$
3 Work out the remaining number. $40 = 40$
4 Add the totals from steps 2 and 3 to find the relative formula mass. $= 164$

Conservation of mass

- In any chemical equation, the total **mass** of the **reactants** equals the total mass of the **products**. This is called **conservation of mass**.

G–E

If magnesium is heated in oxygen, it will make magnesium oxide.

If 2.4 g of magnesium is burned, 4.0 g of magnesium oxide is made. This is because 1.6 g of oxygen combines with the magnesium.

2.4 + 1.6 = 4.0

By ratios, 4.8 g magnesium will burn to give 8 g of magnesium oxide.

- Mass is conserved because **atoms** cannot be created or destroyed, only rearranged into different **compounds**.

D–C

In the reaction $KOH + HNO_3 \rightarrow KNO_3 + H_2O$
the formula masses of the reactants = 119
the formula masses of the products = 119

Remember!
The mass of product is directly proportional to the mass of the limiting reactant used.

Improve your grade

Conservation of mass

If 125 g of zinc carbonate made 81 g of zinc oxide and 44 g of CO_2, show that mass is conserved in this reaction.

$ZnCO_3 \rightarrow ZnO + CO_2$

AO2 [3 marks]

Percentage yield and atom economy

Yield and percentage yield

- The **percentage yield** of a reaction is the amount of **product** made (the actual yield) compared to the amount we expect to make (the predicted yield).
  - 100% yield means that no product has been lost.
  - 0% yield means that no product has been collected.

- If all the lost bits of chemical could be collected, the total **mass** of product would be the same as the total mass of **reactant**. Some of the ways that bits are lost in a typical experiment are:
  - loss in **filtration** – small amounts soak into or stay on the filter paper
  - loss in **evaporation** – some chemicals spit out or evaporate into the room
  - loss in transferring liquids – tiny amounts of liquid stick to the sides of the beaker when the **solution** is poured
  - some of the reactants do not react to form the products.

Calculating percentage yield and industrial costs

- Relative formula mass calculations can be used to find out how much product should be made (the predicted yield) in a chemical reaction.

- In practice, the actual yield produced is less, due to losses in the practical methods used.

- Percentage yield is calculated using the formula:

$$\text{Percentage yield} = \frac{\text{actual yield}}{\text{predicted yield}} \times 100$$

> **EXAM TIP**
>
> You need to learn both of the formulae on this page, and be able to use them.

A soap manufacturer calculates the yield required as 150 tonnes. However, when the soap is made, only 147.6 tonnes have been produced. The percentage yield can be calculated as follows:

$$\text{Percentage yield} = \frac{147.6}{150} \times 100 = 98.4\ \%$$

Atom economy

- **Atom economy** is a way of measuring the amount of **atoms** that are wasted or lost when a chemical is made.
  - 100% atom economy means that all the atoms in a reactant have changed into the desired product.
  - The higher the atom economy, the 'greener' the process.

- Atom economy can be found using the formula:

$$\text{Atom economy} = \frac{M_r \text{ of desired products}}{\text{Sum of } M_r \text{ of all products}} \times 100$$

A company makes potassium nitrate for use as a **fertiliser**, as a rocket propellant and in fireworks. H_2O is a waste product.
The chemical can be made using this reaction:

	KOH	+	HNO$_3$	→	KNO$_3$	+	H$_2$O
M_r	56	+	63	→	101	+	18

$$\text{Atom economy} = \frac{101}{119} \times 100 = 84.9\%$$

> **Remember!**
> When calculating atom economy, use only the relative formula mass of the products.

Improve your grade

Percentage yields and atom economy

A reaction has an atom economy of 86% and a percentage yield of 52%.
What does this mean? *AO2* [2 marks]

Energy

Exothermic and endothermic reactions

- **Exothermic reactions** give out **energy** to the surroundings (release energy).
- **Endothermic reactions** take in energy from the surroundings (absorb energy).
- Measuring **temperature** changes shows the type of reaction.
  - If the temperature goes up, it is an exothermic reaction.
  - If the temperature drops, it is an endothermic reaction.

- Bond breaking is an endothermic process.
- Bond making is an exothermic process.

> **Remember!**
> All chemical reactions involve bond making (exothermic) and bond breaking (endothermic).

G–E

D–C

Comparing the energy in fuels

- The apparatus on the right can be used to compare the energy released by two fuels in **combustion** reactions.
- Each fuel needs to be burnt in a bottled gas burner or spirit burner.
- Water is heated in a copper calorimeter.
- To compare different fuels fairly:
  - use the same size and type of calorimeter
  - measure out the same volume of water
  - use water that starts near to the same temperature
  - adjust the flame so it is the same size
  - ensure the flame is the same distance from the calorimeter
  - burn the fuels for the same amount of time.
- To compare fuels, find out how much the temperature of the water goes up.

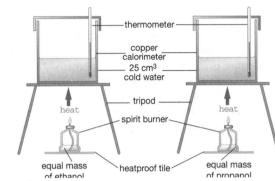

Labels: thermometer; copper calorimeter; 25 cm³ cold water; tripod; heat; spirit burner; equal mass of ethanol; heatproof tile; equal mass of propanol; heat

Measuring the energy transferred by different fuels

G–E

Fuel	Temperature at start (°C)	Temperature at end (°C)	Temperature change (°C)
methanol	25	35	10
ethanol	26	46	20
propanol	24	53	29

- You can see from this experiment that the propanol released more heat energy.
- The energy transferred by a fuel is calculated using the formula:

Energy transferred (in J) = m × c × ΔT

where m = **mass** of water heated (in g)
      c = **specific heat capacity** of water (4.2 J/g °C)
    ΔT = temperature change (in °C).

- To find out the energy released by 1 g of solid fuel:
  - Measure out the mass of 1 g of fuel.
  - Pour 100 g of water into a copper calorimeter (1 cm³ of water = 1 g).
  - Heat the water with the burning fuel.
  - Measure the temperature rise.
  - Ensure reliability by repeating results.

EXAM TIP

When answering calculation questions, remember to show all the stages and include units in your answers.

D–C

Improve your grade

Comparing energy in fuels
Three fuels heat the same mass of water when burnt under a calorimeter. Fuel A changes the temperature by 21 °C, Fuel B by 25 °C and Fuel C by 18 °C. Which fuel released the most energy? *AO2* [1 mark]

Batch or continuous?

Batch and continuous processing

- Chemicals needed in large amounts are usually made by a *continuous process*. Two examples of chemicals made by this method are:
  - ammonia, used to make **fertilisers**
  - sulfuric acid, used to make thousands of different **compounds**.
- To make ammonia, nitrogen and hydrogen are continuously pumped into a reaction vessel and heated under pressure.
- Continuous processing works day and night.
- Chemicals needed in small quantities are made using **batch processing**.
- **Pharmaceuticals** (medical drugs) are examples of chemicals made using the batch process, as they are only made when needed.

- Continuous processing:
  - makes large amounts of the **product** 24 hours a day, seven days a week
  - takes place in large chemical plants with good transport links; plants are highly automated so have minimal labour costs, making the product cheaper
  - takes less energy to maintain – as long as the process can be kept running – than to stop and start the process.
- Batch processing:
  - makes a fixed amount
  - allows batches to be made and stored until needed
  - allows quantities to be made that can be sold within a given time as many drugs have an expiry date
  - makes it easy to make a new batch when needed
  - makes it easy to change production to a different product.

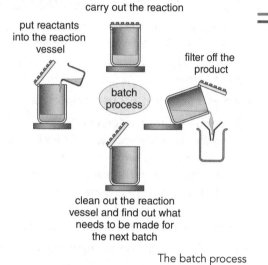

The batch process

Why are medicines so expensive?

- Whichever way medicines are made, there are costs.
- The chemicals in medical drugs need to be as pure as possible. Impurities might give side effects. All new drugs need to be tested to make sure that they work and do not have serious side effects.
- Three tests can be used to find if a chemical compound is pure. A pure chemical will:
  - melt at a fixed temperature (the **melting point**)
  - boil at a fixed temperature (the **boiling point**)
  - give the same result when tested using thin layer **chromatography**.
- The raw materials for medicines can be chemicals that are synthetic or chemicals extracted from plants.

Factors affecting the cost of making a medicine

- Extracting a chemical from a plant involves:
  - crushing to disrupt and break cell walls
  - boiling in a suitable solvent to dissolve the compounds
  - chromatography to separate and identify individual compounds
  - isolating, purifying and testing for potentially useful compounds.
- Drug prices are set to take account of the development costs.
  - It can take about 10 years to develop and test a new drug and each country has strict safety laws that pharmaceutical companies must follow.
  - Many compounds need to be made before one may be found which is useful to develop.
  - Raw materials are often rare and costly.
  - Many raw materials are found in plants, and are difficult to extract.

Improve your grade

Drug development
What are the factors that affect the making of a drug? *AO1* [3 marks]

C3 Chemical economics

Allotropes of carbon and nanochemistry

Allotropes

- Diamond, **graphite** and **buckminsterfullerene** are all made from **carbon**.

- Carbon **atoms** can join with themselves in different ways so that different structures can form.

- The table shows the **physical properties** of diamond and graphite.

part of diamond structure

complete particle of buckminsterfullerene

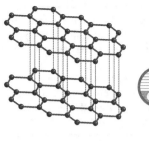
part of graphite structure

Structure of diamond, buckminsterfullerene and and graphite

Diamond	Graphite
lustrous (shiny), transparent and colourless	lustrous, opaque and black
very hard	soft and slippery
very high melting point	very high melting point
insoluble in water	insoluble in water
does not conduct electricity	conducts electricity

- Diamond, graphite and **fullerenes** are all **allotropes** of carbon. Allotropes are different structures of the same **element**.

Nanotubes

- Carbon can also be used to make very small structures called **nanotubes**.

- Nanotubes have special properties and are useful:
  - as semiconductors in electrical circuits
  - as industrial **catalysts**
  - for reinforcing graphite in tennis rackets as they are very strong.

- Fullerenes are carbon structures that form spheres or tubes and can be used:
  - to carry and deliver drugs **molecules** around the body
  - to trap dangerous substances in the body and remove them.

- Buckminsterfullerene contains 60 carbon atoms in a sphere. Its formula is written as C_{60}. Each sphere is so small it is measured in **nanometres**. A nanometre is 10^{-9} metres long.

Carbon nanotubes

Why diamonds and graphite are useful

- Diamond and graphite are both giant covalent structures of carbon atoms.
  - Every carbon atom in diamond makes strong **covalent bonds** in different directions and in graphite the carbon atoms bond to make layers.

- Diamond is the hardest natural substance known and has a very high **melting point** and **boiling point**. The imperfections that naturally occur in diamonds form cleaving plates which allow them to be shaped. These properties make them ideal for use in cutting tools. The imperfections allow it to be shaped so that it can be used for jewellery, as the facets reflect light.

- Graphite has a high melting and boiling point, but the layers can slide over each other. It is used in pencils, and as a high-**temperature** lubricant.

Improve your grade

Structure of allotropes
Give two properties of diamond that are different from the properties of graphite.
AO1 [4 marks]

Rate of reaction is how much product is made in a fixed time.

Fine powders are more combustible as they have a greater surface area exposed.

In a reaction, particles collide. The higher the temperature, the faster the reaction.

Rates of reaction

Increasing pressure increases the rate of reaction.

The higher the concentration, the faster the reaction.

Catalysts change the rate of reaction but remain unchanged at the end.

Rates of reaction can be compared by comparing the gradients of any two lines on the same graph.

The total mass of reactants in a chemical reaction is equal to the total mass of the products: this is called the principle of conservation of mass.

In the reaction:

$KOH + HNO_3 \rightarrow KNO_3 + H_2O$

the total M_r of reactants KOH and HNO_3 =119 and the total M_r of products KNO_3 and H_2O =119, so mass is conserved.

Reacting masses

100% atom economy means that all atoms in the reactant have been converted to the desired product.

The percentage yield can be less than 100% as there is loss in filtration, loss in evaporation, loss in transferring liquids and not all reactants react to make products.

$$\text{Percentage yield} = \frac{\text{actual yield}}{\text{predicted yield}} \times 100$$

$$\text{Atom economy is} = \frac{M_r \text{ of desired products}}{\text{Sum of } M_r \text{ of all products}} \times 100$$

An exothermic reaction is one in which energy is transferred out into the surroundings.

Endothermic reactions and exothermic reactions can be recognised using temperature changes.

Bond making is an exothermic process.

Energy

An endothermic reaction is one in which energy is taken in from the surroundings.

Bond breaking is an endothermic process.

Energy is calculated experimentally by:
Energy = mass of water × c × temperature change, where c is the specific heat capacity of water (4.2 J/g °C).

Continuous processes are used to make chemicals such as ammonia.

Chemicals are extracted from plants by crushing, boiling, dissolving and chromatography.

Industrial processes

The cost of making and developing a pharmaceutical drug includes labour, raw materials, energy, research and testing.

Batch processes are used to make pharmaceutical drugs.

Diamond, graphite and fullerenes are allotropes of carbon.

Allotropes

Diamond is colourless and does not conduct electricity but graphite does and is slippery.

Atomic structure

Atoms

- An **atom** is very small. It has a **nucleus** surrounded by **electrons**. The nucleus is positively **charged**, an electron is negatively charged and an atom is **neutral**.

- The nucleus of an atom is made up of **protons** and **neutrons**. The relative charge and mass of electrons, protons and neutrons are given in the table (right).

	Relative charge	Relative mass
electron	−1	0.0005 (zero)
proton	+1	1
neutron	0	1

- The **atomic number** is the number of protons in an atom.

- The **mass number** is the total number of protons and neutrons in an atom.

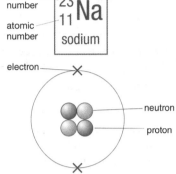

- If a neutral atom has an atomic number of 11 and a mass number of 23 it will be a sodium atom and have:
  - 11 protons (because of its atomic number) and 11 electrons
  - 12 neutrons, making a mass number of 23 (11 protons + 12 neutrons).

- You can deduce the numbers of protons, electrons and neutrons from symbols.

	Atomic number	Mass number	Number of protons	Number of electrons	Number of neutrons
sodium	11	23	11	11	12
$^{19}_{9}F$	9	19	9	9	10
carbon-12	6	12	6	6	6

A helium atom has two protons and two electrons, so the atom is neutral because it has the same number of protons as electrons; the charge of the neutron is 0

Isotopes

- **Elements** with the *same* atomic number but *different* mass numbers are **isotopes**.

- $^{12}_{6}C$ has six protons and six neutrons. $^{14}_{6}C$ has six protons and a mass of 14; it has eight neutrons (14 − 6). These are both isotopes. $^{14}_{6}C$ can also be written as **carbon-14**.

Isotope	Electrons	Protons	Neutrons
$^{12}_{6}C$	6	6	6
$^{14}_{6}C$	6	6	6

Arrangement of electrons in atoms

- Elements have atoms that are all the same. Each element is identified in the **periodic table**.

- Each element has a symbol. C is **carbon**, Co is cobalt, Cl is chlorine and Cu is copper. If two capital letters appear it means that it is a formula of a **compound**. For example, CO_2 is **carbon dioxide**.

- The elements in the periodic table are arranged in order of increasing atomic number.
  - An element with an **electronic structure** of 2.8.6 has three **electron shells**, so is in the third row.
  - The electronic structure 2.8.6 means the atomic number is 16, so the element is sulfur.

- Elements in the same *group* (with the same number of electrons in the outer shell) are arranged vertically ↑.

- Elements in the same **period** (in order of how many shells the electrons occupy) are arranged horizontally →.

The development of atomic theory

- John Dalton developed an early theory about atoms. When J.J. Thompson discovered the electron, Rutherford discovered the nucleus and Bohr found new evidence for electron orbits, their explanations *changed* the model of the atom.

- John Dalton's explanation was *provisional*, but was confirmed by better *evidence* from the other scientists later. When J.J. Thompson, Rutherford and Bohr found new evidence, their explanations changed the model of the atom. With later evidence their *predictions* were confirmed.

Improve your grade

Arrangement of electrons

If a potassium atom has the electronic structure 2.8.8.1, how many shells do the electrons occupy? *AO2* [1 mark]

Ionic bonding

Forming ions and molecules

- Li is the symbol of a lithium **atom**.

- A **molecule** has more than one atom in its formula and no **charge**. CO_2 (**carbon dioxide**) is a molecule.

- An **ion** is a charged atom or group of atoms and can be recognised from its formula. Li^+ is an ion.
  - A calcium ion has two positive charges. It is a **positive ion** (Ca^{2+}).
  - Ions can have a positive charge, such as Ca^{2+}, or they can have a negative charge, such as Cl^-. This is a **negative ion**.
  - Ions can also be groups of atoms with a charge. For example, NH_4^+ or NO_3^-.

Why do atoms form bonds?

- Atoms with an outer shell of eight **electrons** have a **stable electronic structure**.

- Atoms can be made stable by transferring electrons. This called **ionic bonding.**

- **Metal** atoms lose electrons to get a stable electronic structure. If an atom loses electrons, a positive ion is formed. This is because there are fewer negatively charged electrons than the number of positively charged **protons** in the **nucleus**.

- **Non-metal** atoms gain electrons to get a stable electronic structure. If an atom gains electrons, a negative ion is formed. This is because there will be more negatively charged electrons than the number of positively charged protons in the nucleus.

- During ionic bonding, the metal atom becomes a positive ion and the non-metal atom becomes a negative ion. The positive ion and the negative ion then **attract** one another.

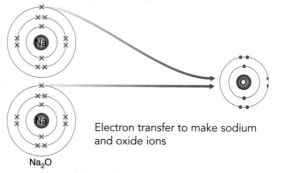

Electron transfer to make sodium and oxide ions

Na_2O

Conducting electricity

- Solid sodium chloride does not conduct electricity. **Molten** sodium chloride and sodium chloride **solution** both do.

- Both sodium chloride and magnesium oxide have high **melting points**, but magnesium oxide has a higher melting point than sodium chloride.

- The structure of sodium chloride or magnesium oxide is a **giant ionic lattice**. The positive ions have strong **electrostatic attraction** to negative ions. They always exist as solids.

The structure of sodium chloride

- Both sodium chloride and magnesium oxide conduct electricity when they are molten.

- Sodium chloride solution can conduct electricity.

Improve your grade

Structure and bonding

Sodium chloride is made up of Na^+ particles and Cl^- particles. Are these atoms, ions or molecules? Is sodium chloride easier or harder to melt than magnesium oxide?
AO2 [2 marks]

C4 The periodic table

The periodic table and covalent bonding

Covalent bonding

- There are two types of bonding:
  - **ionic bonding** between **metals** and **non-metals**
  - **covalent bonding** between non-metals.
- Non-metals can share **electron** pairs between **atoms** to make a covalent bond.

CO_2 is covalently bonded

A molecule of carbon dioxide

Predicting chemical properties

- **Carbon dioxide** and water do not conduct electricity.
- The **attraction** between carbon dioxide **molecules** is called an **intermolecular force**. The attraction between water molecules is also an intermolecular force.

weak intermolecular forces between the molecules

Intermolecular forces between molecules

Group numbers and elements in periods

- A *group* of **elements** is all the elements in a vertical column of the **periodic table**. These elements have similar **chemical properties**.
- A **period** of elements is all the elements in a horizontal row of the periodic table.
- The group number is the same as the number of electrons in the outer shell.
- If an atom of an element has electrons in only one occupied shell it will be found in the first period; if it has electrons in two occupied shells it will be in the second period; if it has electrons in three occupied shells, it will be found in the third, and so on.

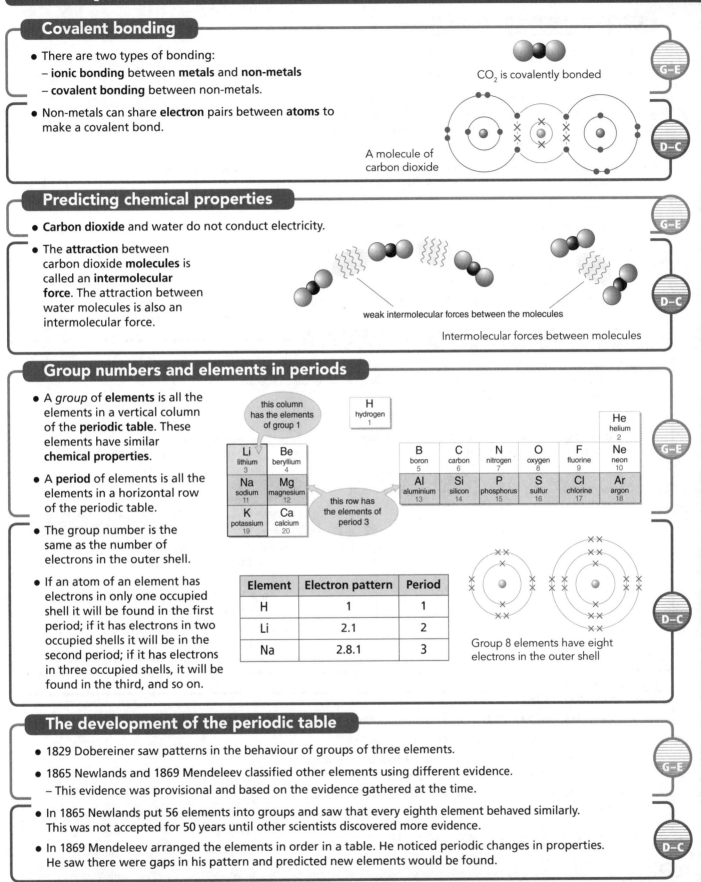

this column has the elements of group 1

this row has the elements of period 3

Group 8 elements have eight electrons in the outer shell

Element	Electron pattern	Period
H	1	1
Li	2.1	2
Na	2.8.1	3

The development of the periodic table

- 1829 Dobereiner saw patterns in the behaviour of groups of three elements.
- 1865 Newlands and 1869 Mendeleev classified other elements using different evidence.
  - This evidence was provisional and based on the evidence gathered at the time.
- In 1865 Newlands put 56 elements into groups and saw that every eighth element behaved similarly. This was not accepted for 50 years until other scientists discovered more evidence.
- In 1869 Mendeleev arranged the elements in order in a table. He noticed periodic changes in properties. He saw there were gaps in his pattern and predicted new elements would be found.

Improve your grade

Elements in the periodic table

Chlorine is number 17 in the periodic table. Name two other elements that are in the same group as chlorine and two other elements that are in the same period.
AO2 [2 marks]

The group 1 elements

Properties of alkali metals

G–E

- Lithium, sodium and potassium are **group 1 metals**.

- They are stored under oil as they react with both water and air.

- The **metals** in this group are known as the **alkali metals**.
  - They are called 'alkali metals' because they react with water to give alkaline **solutions** (metal hydroxides). The reaction is very quick and the metals float.
  - Hydrogen is also formed in this reaction.

- The reactivity of group 1 metals with water is vigorous and increases down the group.
  - The word equation for one of these reactions is:

    potassium + water → potassium hydroxide + hydrogen

  - Potassium is very reactive and gives a lilac flame in a flame test.

Remember!
Group 1 metals are called alkali metals because they react with water to form alkalis.

D–C

- Caesium and rubidium are both group 1 metals and so have similar properties to lithium, sodium and potassium:
  - They react vigorously with water.
  - Hydrogen gas is given off.
  - The metal reacts with water to form an **alkali** – the hydroxide of the metal.

Predicting the properties of alkali metals

D–C

- It is possible to predict how an alkali metal, such as caesium and rubidium, will behave by looking at the pattern of reactivity in other alkali metals.
  - Sodium reacts more vigorously than lithium.
  - Potassium reacts more vigorously than sodium.

- **Atoms** of group 1 alkali metals have similar properties because they have one **electron** in their outer shell.

Sodium has one electron in its outer shell

reactivity increases down the group

	Melting point in °C	Boiling point in °C
$_3$Li	179	1317
$_{11}$Na	98	892
$_{19}$K	64	774

Reactivity of the alkali metals with water increases down group 1

- The **balanced symbol equation** for lithium and water is:

$$2Li + 2H_2O \rightarrow 2LiOH + H_2$$

EXAM TIP

Construct balanced symbol equations in steps.
Step 1: Write a word equation.
lithium + water → lithium hydroxide + hydrogen
Step 2: Put in the symbols and formulae.
$Li + H_2O \rightarrow LiOH + H_2$
Step 3: Balance up the number of atoms on either side.
$2Li + 2H_2O \rightarrow 2LiOH + H_2$

Flame tests

D–C

- A **flame test** can be used to find out if lithium, sodium or potassium are present in a **compound**.

- The test is carried out while wearing safety glasses.
  - A flame-test wire is moistened with dilute hydrochloric acid.
  - The flame-test wire is dipped into the solid chemical.
  - The flame-test wire is put in a blue Bunsen burner flame.
  - The colours of the flames are recorded.

- The flame test colours are shown in the table.

G–E

Alkali metal in the compound	Colour of flame
lithium	red
sodium	yellow
potassium	lilac

Improve your grade

Reactivity in alkali metals
Explain how group 1 elements react with water. AO1 [4 marks]

The group 7 elements

Halogens

- Fluorine, chlorine, **bromine** and **iodine** are **group 7 elements**.
  - Chlorine is used to sterilise water, make pesticides and make plastics.
  - Iodine is used to sterilise wounds.
- Group 7 **elements** are known as the **halogens**.

| F |
| fluorine |
| 9 |
| Cl |
| chlorine |
| 17 |
| Br |
| bromine |
| 35 |
| I |
| iodine |
| 53 |

The halogens in the periodic table

- At room temperature, *chlorine* is a green gas, *bromine* is an orange liquid and *iodine* is a grey solid.
- Group 7 elements have similar properties as their **atoms** have *seven* **electrons** in their outer shell.

Electronic structures of the halogens

fluorine 2. 7   chlorine 2. 8. 7   bromine (outer shell only shown) 7   iodine (outer shell only shown) 7

The reactions of halogens

- Group 7 elements react vigorously with group 1 elements.
- If the **product** of a reaction between a group 7 and a group 1 element is known, such as potassium iodide, it is possible to write a word equation for the reaction.

potassium + iodine → potassium iodide

- When halogens react with **alkali metals**, a **metal halide** is made. For example, when potassium reacts with iodine, the metal halide made is potassium iodide.
- The word equation for a reaction of a group 1 element and a group 7 element can be worked out from the two **reactants**, as in the reaction between potassium and iodine, above.
- The **balanced symbol equation** for this reaction is:

$2K + I_2 \rightarrow 2KI$

Displacement reactions of halogens

- The reactivity of group 7 elements decreases down the group.
  - Chlorine is less reactive than fluorine, bromine is less reactive than chlorine. Iodine is less reactive than bromine.
  - The reaction between chlorine and potassium bromide produces potassium chloride and bromine. The word equation for this reaction is:

chlorine + potassium bromide → potassium chloride + bromine (orange solution)

- If halogens are bubbled through **solutions** of metal halides there are two possible outcomes: no reaction or a **displacement reaction**. Two displacement reactions are shown in the examples below.
  - Chlorine displaces the bromide to form bromine solution.

$Cl_2 + 2KBr \rightarrow 2KCl + Br_2$ (balanced)

  - Bromine displaces iodides from solutions.

$Br_2 + 2KI \rightarrow 2KBr + I_2$ (red–brown solution)

Remember!
The '2' in Cl_2 means there are two bonded atoms in the **molecule** of chlorine. This number cannot be changed!

Improve your grade

Displacement reactions of halogens

If chorine is bubbled into a solution of potassium iodide, iodine and potassium chloride are made. Construct a word equation for this reaction. *AO1* [2 marks]

Transition elements

Coloured compounds

- The **elements** Sc to Zn are the top row of the **transition elements** in the **periodic table**. The full table, showing their position, is on page 248.
  - The periodic table can be used to find the name or symbol of a transition element.
  - The transition elements include iron (Fe) and copper (Cu).

45 Sc scandium 21	48 Ti titanium 22	51 V vanadium 23	52 Cr chromium 24	55 Mn manganese 25	56 Fe iron 26	59 Co cobalt 27	59 Ni nickel 28	63.5 Cu copper 29	65 Zn zinc 30
89 Y yttrium 39	91 Zr zirconium 40	93 Nb niobium 41	96 Mo molybdenum 42	[98] Tc technetium 43	101 Ru ruthenium 44	103 Rh rhodium 45	106 Pd palladium 46	108 Ag silver 47	112 Cd cadmium 48
139 La* lanthanum 57	178 Hf hafnium 72	181 Ta tantalum 73	184 W tungsten 74	186 Re rhenium 75	190 Os osmium 76	192 Ir iridium 77	195 Pt platinum 78	197 Au gold 79	201 Hg mercury 80

The transition elements

- Transition elements are **metals**, with **metallic properties**. They:
  - conduct heat
  - are shiny (lustrous)
  - conduct electricity
  - are sonorous (ring when struck)
  - are malleable (can be beaten into a sheet)
  - are ductile (can be drawn into a wire).

- **Compounds** that contain a transition element are often coloured.
  - Copper compounds are often blue.
  - Iron(II) compounds are often pale green.
  - Iron(III) compounds are often orange/brown.

> **Remember!**
> A catalyst is a chemical that speeds up a reaction but is not changed or used up by the reaction.

Catalysts

- Transition elements and their compounds are often used as **catalysts**.
  - Iron is used in the **Haber process** to make ammonia, which is used in **fertilisers**.
  - Nickel is used in the manufacture of margarine to harden the oils.

Thermal decomposition of metal carbonates

- **Thermal decomposition** is when a substance is broken down into at least two other substances by heat.
  - If **carbon dioxide** is given off during this type of reaction, it can be tested as it turns limewater milky.

- If a transition metal carbonate is heated, it undergoes thermal decomposition to a metal *oxide* and carbon dioxide. For example, the word equation for the decomposition of copper carbonate to copper oxide and carbon dioxide is:

  copper carbonate → copper oxide + carbon dioxide

- If a transition metal carbonate is heated it undergoes thermal decomposition to form a metal oxide and carbon dioxide. The model of this reaction is shown on the right. On heating:
  - $FeCO_3$ decomposes to form iron oxide and carbon dioxide
  - $CuCO_3$ decomposes to form copper oxide and carbon dioxide
  - $MnCO_3$ decomposes to form manganese oxide and carbon dioxide
  - $ZnCO_3$ decomposes to form zinc oxide and carbon dioxide.

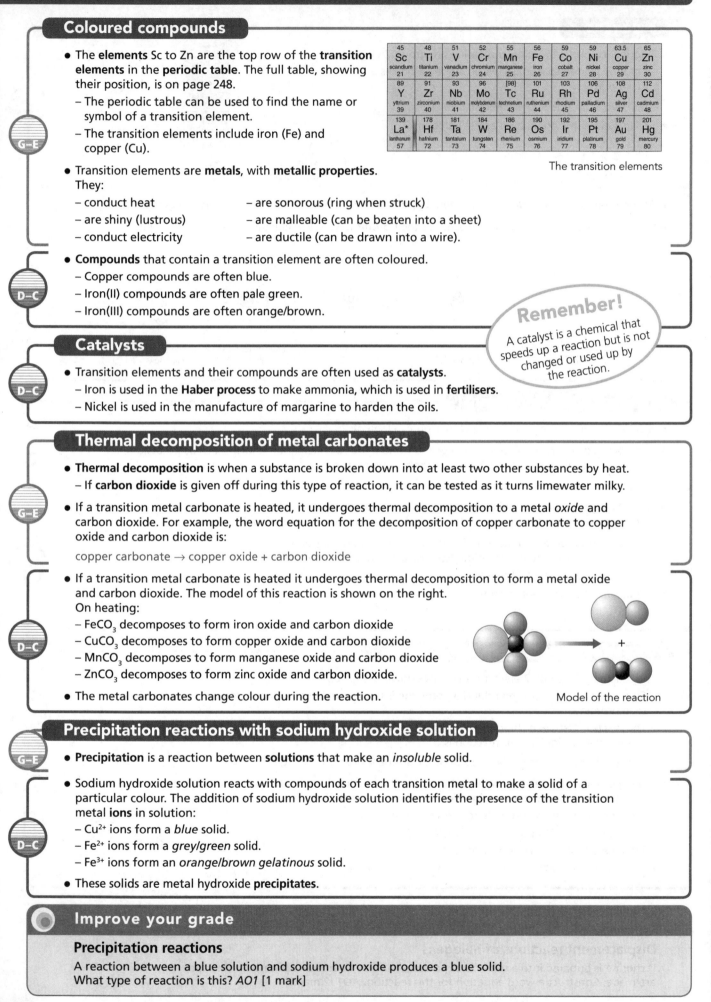

Model of the reaction

- The metal carbonates change colour during the reaction.

Precipitation reactions with sodium hydroxide solution

- **Precipitation** is a reaction between **solutions** that make an *insoluble* solid.

- Sodium hydroxide solution reacts with compounds of each transition metal to make a solid of a particular colour. The addition of sodium hydroxide solution identifies the presence of the transition metal **ions** in solution:
  - Cu^{2+} ions form a *blue* solid.
  - Fe^{2+} ions form a *grey/green* solid.
  - Fe^{3+} ions form an *orange/brown gelatinous* solid.
- These solids are metal hydroxide **precipitates**.

Improve your grade

Precipitation reactions

A reaction between a blue solution and sodium hydroxide produces a blue solid.
What type of reaction is this? *AO1* [1 mark]

Metal structure and properties

Properties of metals

- **Metals** have a number of particular properties, they:
  - are good *conductors* of heat and electricity
  - are shiny (lustrous) and hard
  - have a high tensile strength and high **density**
  - have high **melting** and **boiling points**.

- These properties make metals good for some uses.
  - Saucepan bases need to be made from good conductors of heat.
  - Bridges and cars are made from iron as it is strong.
  - Electrical wiring is made from copper as it is a good **electrical conductor**.

- A property can be either *physical* or *chemical*.
  - An example of a **physical property** is the high *thermal conductivity* of copper. Saucepan bases need to be good conductors of heat, so copper can be used for the base or it can be used for the whole of the pan.
  - An example of a **chemical property** is the *resistance to attack* by oxygen or acids. This is shown by gold. Copper is also resistant, which is another reason why it is used for saucepans.
  - Other physical properties of metals include being lustrous, **malleable** or *ductile*.
  - Aluminium has a low density and is used where this property is important, such as in the aircraft industry and also in modern cars.

Metallic bonding

- Particles in metals are held together by metallic bonds.

- Metals have high melting points and high boiling points, due to their strong **metallic bonds**. The bonds between these **atoms** are very hard to break. A lot of energy is needed to break the atoms apart.

- When metals conduct electricity, **electrons** in the metal move.
  - Copper, silver and gold conduct electricity very well.

electrons from outer shells of metal atoms are free to migrate

The structure of a metal

metal ions

Superconductors

- At very low **temperatures** some metals can be **superconductors**.

- Superconductors are materials that conduct electricity with little or no **resistance**.
  - Copper, silver and gold conduct electricity well, but surprisingly do not become superconductors.

- The electrical resistance of mercury suddenly drops to zero at −268.8 °C. This phenomenon is called superconductivity.

- When a substance goes from its normal state to a superconducting state, it no longer has any magnetic fields inside it.
  - If a small magnet is brought near the superconductor, it is **repelled**.
  - If a small permanent magnet is placed above the superconductor, it levitates.

- The potential benefits of superconductors are:
  - loss-free **power transmission**
  - super-fast electronic circuits
  - powerful **electromagnets**.

Improve your grade

Uses of metals
Explain which properties of metals are useful when making gardening tools. *AO2* [2 marks]

Purifying and testing water

Water purification

- Different types of water resources are found in the UK:
  - lakes and rivers
  - *aquifers* and **reservoirs**.
- Water is an important resource for many industrial chemical processes, either as a raw material or as a coolant.
- Domestic water supplies have to be purified as they may contain substances such as:
  - dissolved salts and minerals, insoluble materials and **microbes**
  - **pollutants**, which can be *nitrate residues*, *lead compounds* and **pesticide residues**.
- Microbes can be killed by **chlorination**.

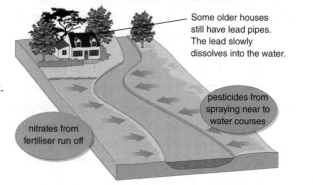

Some older houses still have lead pipes. The lead slowly dissolves into the water.

pesticides from spraying near to water courses

nitrates from fertiliser run off

The pollutants that can get into drinking water

- The water in a river is cloudy and often not fit to drink.
- To turn it into the clean water in taps it is passed through a water purification works.
- Some pollutants, such as nitrates from **fertilisers** and pesticides from crop spraying, get into the water before purification, and some, such as **lead** from old water pipes, get in after treatment.
- There are three main stages in water purification:
  - **sedimentation** – chemicals are added to make solid particles and **bacteria** settle out
  - **filtration** of very fine particles – a layer of sand on gravel filters out the remaining fine particles; some types of sand filter also remove microbes
  - *chlorination* – chlorine is added to kill microbes.
- These steps cost money so water must be conserved to provide clean drinking water.
  - Water is a *renewable* resource, but the supply is not endless. If there is not enough rain in the winter, reservoirs do not fill up.
  - In the UK more homes are being built, increasing the demand for water.
  - It takes **energy** to pump and purify, which increases **global warming**.

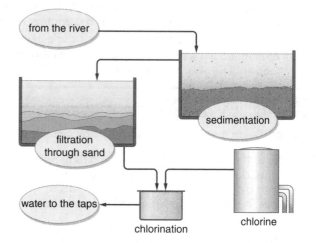

from the river

sedimentation

filtration through sand

water to the taps

chlorination

chlorine

The steps in the purification of water

Remember!
Sulfates are tested by barium chloride. Halides are tested by silver nitrate.

Water tests

- Water can be tested with **precipitation reactions** using aqueous **silver nitrate** and **barium chloride** solutions.
  - Barium chloride **solution** is used to test for sulfate **ions**. Sulfates give a white **precipitate**.
  - For example, the reaction between barium chloride and magnesium sulfate to create barium sulfate and magnesium chloride is:

  barium chloride + magnesium sulfate → barium sulfate (white precipitate) + magnesium chloride

  - Silver nitrate solution makes precipitates with chloride ions, bromide ions and iodide ions.

Halide	Chloride	Bromide	Iodide
colour of precipitate	white	cream	yellow

- In a precipitation reaction, two solutions react to form a chemical that does not dissolve. This chemical suddenly appears in the liquid as a solid – a precipitate.

  silver nitrate + sodium bromide → silver bromide (cream precipitate) + sodium nitrate

  silver nitrate + sodium iodide → silver iodide (yellow precipitate) + sodium nitrate

  silver nitrate + sodium chloride → silver chloride (white precipitate) + sodium nitrate

Improve your grade

Testing water

Explain how you would test whether the ions in a sample of water were chloride or bromide ions. *AO1* [3 marks]

C4 Summary

The nucleus is made up of protons and neutrons.

The electron has a charge of −1, the proton has a charge of +1, the neutron has a charge of 0.

Metals form positive ions as they lose electrons.

Non-metals form negative ions as they gain electrons.

Isotopes are elements that have atoms of the same atomic number but different mass numbers.

Atomic structure and atomic bonding

Ionic bonding happens because of the transfer of electrons from metals to non-metals.

Sodium chloride and magnesium oxide form giant ionic lattices as their ions attract.

Magnesium oxide and sodium chloride conduct electricity when molten.

Non-metals combine by sharing electrons.

Carbon dioxide and water do not conduct electricity.

Carbon dioxide and water are simple molecules with weak intermolecular forces between the molecules.

The periodic table and covalent bonding

A period of elements is all the elements that are in the same horizontal row in the periodic table.

A group of elements is all the elements in a vertical column of the periodic table, which all have similar chemical properties.

The period to which an element belongs is the same as the number of occupied electron shells.

The group number is the same as the number of electrons in the outside shell.

Rubidium and caesium are group 1 metals which react violently with water to give off hydrogen and make an alkaline solution.

Group 1 metals have similar properties as they all need to lose one electron from their outside shell.

Lithium compounds give a red flame, sodium compounds give yellow and potassium compounds give a lilac flame.

Groups

The reactivity of group 7 elements decreases down the group.

Flame tests can be used to identify the presence of lithium, sodium and potassium in compounds.

Chlorine is a green gas, bromine is a brown liquid and iodine is a grey solid.

If a group 1 metal reacts with a group 7 non-metal the word equation for the formation of a metal halide can be constructed.

Copper compounds are often blue, iron(II) compounds are often light green and iron(III) compounds are often orange/ brown.

The thermal decomposition of transition metal carbonates results in the metal oxide and carbon dioxide being made.

Carbon dioxide turns limewater milky.

Cu^{2+} ions react with sodium hydroxide to make a blue solid in a precipitation reaction.

Metals are lustrous, hard, have high tensile strength and are good conductors of electricity.

Metals and water

Metals have high melting points due their strong metallic bonding.

At low temperatures some metals can be superconductors, which conduct electricity with little or no resistance.

Drinking water is purified by filtration, sedimentation and chlorination.

Chlorination kills microbes in water.

Sulfates in water can be tested using barium chloride, halides can be tested using silver nitrate.

Speed

Measuring speed

- **Speed** is a measure of how fast something is going.
- To determine speed you need to measure distance and time. Faster objects cover more distance in a given time.
- The unit of speed is metres per second (m/s) or kilometres per hour (km/h). So, if a car is travelling at 80 km/h it goes a distance of 80 km in a time of 1 hour.
- In everyday situations a tape measure or trundle wheel and stopwatch can be used to measure speed. On roads, a speed camera takes a photograph as a car passes. A second photograph is taken 0.5 seconds later. There are white lines painted on the road 1.5 m apart which show how far the car travels in 0.5 seconds.

G–E

- A car passing over six lines travels:

  1.5 × 6 = 9 m in 0.5 seconds
- This means the **average speed** of the car was 18 m/s or 65 km/h.
- The formula for speed is:

  $\text{average speed} = \dfrac{\text{distance}}{\text{time}}$
- We write 'average speed' because the speed of a car changes during a journey.

An aircraft travels 1800 km in 2 hours.

Average speed = $\dfrac{1800}{2}$ = 900 km/h = 900 × $\dfrac{1000}{3600}$ = 250 m/s

> **Remember!**
> There are 3600 seconds in one hour.

D–C

- The speed of a car at a certain point in time is called **instantaneous speed**.

  distance travelled = average speed × time

  $= \dfrac{(u + v)t}{2}$

> **Remember!**
> You may be asked to rearrange equations. Practise doing this.

- Increasing the speed means increasing the distance travelled in the same time. Increasing the speed reduces the time needed to cover the same distance.

Distance–time graphs

- Drawing a graph of distance against time shows how the distance moved by an object from its starting point changes over time.

G–E

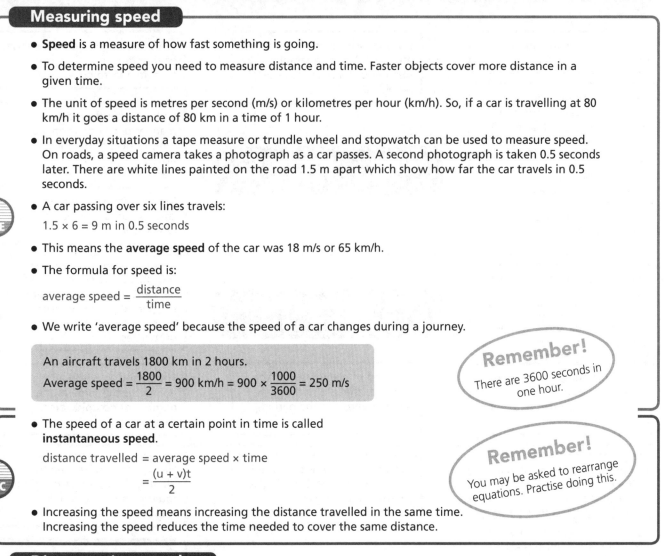

Distance–time graphs

- In graph **a**, the distance doesn't change over time, so the car is stationary. In graph **b**, the distance travelled by the car increases at a steady rate. It's travelling at a constant speed.

D–C

- The **gradient** of a **distance–time graph** tells you about the speed of the object. A higher speed means a steeper gradient. In graph **b**, the distance travelled by the object each second is the same. The gradient is constant, so the speed is constant. In graph **c**, the distance travelled by the object each second increases as the time increases. The gradient increases, so there is an increase in the speed of the object.

Improve your grade

Calculating speed

A car travels 600 m in 20 s. Calculate its average speed.
AO1 [2 marks], *AO2* [1 mark]

Changing speed

Speed–time graphs

- A **speed–time graph** shows how the **speed** of an object, for example a car, changes with time.

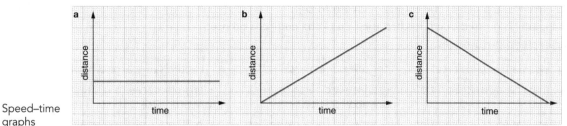

Speed–time graphs

- The **gradient** (slope) of a line tells us how the speed is changing. In graph **a**, the line is horizontal so the speed is constant. In graph **b**, the line has a positive gradient so the speed is increasing. In graph **c**, the line has a negative gradient so the speed is decreasing.

- If the speed is increasing, the object is **accelerating**. If the speed is decreasing, the object is **decelerating**.

G–E

- The area under a speed–time graph is equal to the *distance* travelled.
  - The speed of car B in the graph is increasing more rapidly than the speed of car A, so car B is travelling further than car A in the same time.
  - The area under line B is greater than the area under line A for the same time.
  - The speed of car D is decreasing more rapidly than the speed of car C, so car D isn't travelling as far as car C in the same time.
  - The area under line D is smaller than the area under line C for the same time.

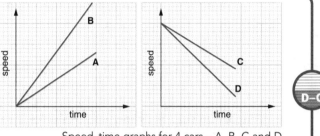

Speed–time graphs for 4 cars – A, B, C and D

D–C

EXAM TIP
Don't confuse distance–time and speed–time graphs. Always look at the axes carefully.

Acceleration

- A change of speed is called acceleration, measured in metres per second squared (m/s²).

- The formula for measuring acceleration is:

$$\text{acceleration} = \frac{\text{change in speed}}{\text{time taken}}$$

- A negative acceleration shows the car is decelerating.

G–E

A new car boasts a rapid acceleration of 0 to 108 km/h in 6 seconds.

A speed of 108 km/h is $\dfrac{108 \times 1000}{60 \times 60} = 30$ m/s

$\text{Acceleration} = \dfrac{\text{change in speed}}{\text{time taken}} = \dfrac{(30 - 0)}{6} = 5$ m/s²

This means the speed of the car increases by 5 m/s every second.

Vectors

- When an object moves, it not only has a speed but also a direction.

- **Velocity** is a measure of speed in a particular direction.

G–E

- When two cars are moving past each other, their **relative velocity** is:
  - the sum of their individual velocities if they are going in opposite directions
  - the difference of their individual velocities if they are going in the same direction.

D–C

Improve your grade

Calculating acceleration
A cat is walking at a velocity of 0.5 m/s when it sees a mouse. The cat takes 2 s to accelerate to 5.5 m/s. Calculate its acceleration in m/s². *AO1* [1 mark] *AO2* [2 marks]

Forces and motion

What do forces do?

- To **accelerate** in a car, the driver presses on the accelerator pedal.
  - This increases the pull of the engine (forward **force**).
  - If the pedal is pressed down further, the pull of the engine is greater.
  - The acceleration increases.

for the same forward force:	for the same **mass**:	for the same acceleration:
more mass has less acceleration	more forward force causes more acceleration	a large mass needs a large forward force
less mass has more acceleration	less forward force causes less acceleration	a small mass needs a small forward force

G–E

Force, mass and acceleration

- If the forces acting on an object are balanced, it's at rest or has a constant **speed**. If the forces acting on an object are unbalanced, it speeds up or slows down.

- The unit of force is the **newton** (N).

force = mass × acceleration

Where F = **unbalanced force** in N, m = mass in kg and a = acceleration in m/s^2.

> Marie pulls a sledge of mass 5 kg with an acceleration of 2 m/s^2 in the snow. The force needed to do this is: F = ma = 5 × 2 = 10 N

Remember! You may be asked to rearrange equations.

D–C

Car safety

- A car driver cannot stop a car immediately. It takes the driver time to react to danger.

- **Thinking distance** is the distance travelled between a driver seeing a danger and taking action to avoid it.

- **Braking distance** is the distance travelled before a car comes to a stop after the brakes have been applied.

- The formula for **stopping distance** is:

stopping distance = thinking distance + braking distance

- A vehicle has to be able to stop safely without hitting anything else.

G–E

- **Reaction time**, and therefore thinking distance, may increase if a driver is:
  - tired
  - under the influence of alcohol or other drugs
  - travelling faster
  - distracted or lacks concentration.

- Braking distance may increase if:
  - the road conditions are poor e.g. icy
  - the car has not been properly maintained e.g. worn brakes
  - the speed is increased.

- For safe driving, it is important to be able to stop safely:
  - Keep an appropriate distance from the car in front.
  - Have different speed limits for different types of road and locations.
  - Slow down when road conditions are poor.

D–C

Improve your grade

Stopping distance

Tim has a reaction time of 0.7 s. Calculate his thinking distance when he is travelling at 108 km/h.
AO1 [1 mark], *AO2* [2 marks]

Work and power

Work

- *Work* is done when a **force** moves. For example, when a person climbs stairs, the force moved is their **weight**.
- The amount of **work done** depends on:
  - the size of the force acting on an object
  - the distance the object is moved.
- **Energy** is transferred when work is done. The more work is done, the more energy is transferred.
- Work and energy are measured in **joules** (J).
- Work is done when a force moves an object in the direction in which the force acts.
- The formula for work done is:

  work done = force × distance moved (in the direction of the force)

  If a person weighs 700 N, the work he does against **gravity** when he jumps 0.8 m is:
  work done = force × distance moved = 700 × 0.8 = 560 J

Remember!
You may be asked to rearrange equations.

G–E

Weight

- Weight is a measure of the gravitational attraction on a body acting towards the centre of the Earth.
- The formula for weight is:

  weight = mass × **gravitational field strength**

- A mass of 1 kg has a weight of about 10 N on Earth.

D–C

Power

- **Power** is the rate at which work is done.
- If two lifts, one old and one new, do the same amount of work but the new one does it more quickly, the new lift has a greater power.
- Power is measured in **watts** (W). A large amount of power is measured in **kilowatts** (kW).
  1 kW = 1000 W.
- The formula for power is:

  $$power = \frac{work\ done}{time\ taken}$$

- A person's power is greater when they run than when they walk.
- Some cars are more powerful than others. They travel faster and cover the same distance in a shorter time and require more fuel. The power rating of a car depends on its engine size. More powerful cars have greater **fuel consumption**.
- Fuel is expensive and a car with high fuel consumption is expensive to run.
- Fuel **pollutes** the environment.
  - Car **exhaust gases**, especially **carbon dioxide**, are harmful.
  - Carbon dioxide is also a major source of **greenhouse gases**, which contribute to climate change.

G–E

D–C

Improve your grade

Comparing power
Jan and Meera can both run up a flight of stairs in 8 seconds. Meera is more powerful than Jan. Explain why. *AO1* [1 mark]

Energy on the move

Kinetic energy

G–E

- Moving objects have **kinetic energy**. Different fuels can be used to gain kinetic energy.
  - The fuel for a car is **petrol** or diesel oil.
- Kinetic energy increases with:
  - increasing **mass**
  - increasing **speed**.

D–C

- The formula for kinetic energy is:
  kinetic energy = $\frac{1}{2}$ mv², where m = mass in kg, v = **velocity** in m/s

Fuel

G–E

- Petrol and diesel oil are **fossil fuels** made from crude oil. Petrol is more refined than diesel oil.
- Petrol cars and diesel cars need different engines. The same amount of diesel oil contains more **energy** than petrol. A diesel engine is more **efficient** than a petrol engine. Diesel cars produce less CO_2 than petrol cars but more nitrogen oxides than petrol cars.
- Some cars use more petrol or diesel oil than others and:
  - cause more **pollution**
  - cost more to run
  - decrease supplies of non-renewable fossil fuels.
- **Biofuels** and *solar energy* are alternative energy sources for vehicles.

D–C

- **Fuel consumption** data are based on ideal road conditions for a car driven at a steady speed in urban and non-urban conditions.

car	fuel	engine in litres	miles per gallon (mpg)	
			urban	non-urban
Renault Megane	petrol	2.0	25	32
Land Rover	petrol	4.2	14	24

EXAM TIP

Make sure you can interpret fuel consumption tables.

Electrically powered cars

D–C

- Electric cars are **battery**-driven or **solar-powered**.
  - The battery takes up a lot of room.
  - They have limited range before **recharging**.
  - They are expensive to buy but the cost of recharging is low.
  - Solar-powered cars rely on the sun shining and need back-up batteries.
- Exhaust fumes from petrol-fuelled and diesel-fuelled cars cause serious pollution in towns and cities.
- Battery-driven cars do not **pollute** the local environment, but their batteries need to be recharged. Recharging uses electricity from a **power station**. Power stations pollute the local atmosphere and cause acid rain.

Streamlining

G–E

- The shape of a vehicle can change its top speed and fuel consumption.
  - Wedge-shaped cars and deflectors on lorries and caravans increase the top speed and reduce fuel consumption.
  - Car roof racks and open windows reduce the top speed and increase fuel consumption.

Improve your grade

Electric cars and pollution

Janice says that electric cars are better for the environment because they don't pollute it. James says that there is still pollution but not at the point of use.
Explain what James means. AO1 [2 marks]

Crumple zones

Momentum and force

- A moving car has **momentum**.

- The formula for momentum is:

  momentum = mass × **velocity** and the units are kgm/s

- If a car is involved in a collision it loses its momentum very quickly.

- To reduce injuries in a collision, forces should be as small as possible.

  $$force = \frac{change\ in\ momentum}{time}$$

- Spreading the momentum change over a longer time reduces the force.

Car safety features

- Modern cars have safety features that absorb **energy** when a vehicle stops suddenly. These are:
  - brakes that get hot
  - **crumple zones** that change shape
  - **seat belts** that stretch a little
  - **air bags** that inflate and squash.
  - collapsible steering column.

- Some safety features are designed to prevent crashes. These include:
  - **ABS brakes** which give stability and maintain steering during hard braking. The driver gets the maximum braking force without skidding and can still steer the car.
  - **Traction control** which stops the wheels on a vehicle from spinning during rapid **acceleration**.
  - Electric windows.
  - **Paddle shift controls** on the steering column.

- On impact:
  - Crumple zones at the front and rear of the car absorb some of its energy by changing shape or 'crumpling'.
  - Seat belts stretch a little so that some of the person's **kinetic energy** is converted to elastic energy.
  - Air bags absorb some of the person's kinetic energy by squashing up around them.

- All these safety features:
  - absorb energy
  - change shape
  - reduce injuries
  - reduce momentum to zero slowly, therefore reducing the force on the occupants.

How does an air bag and a seat belt help to prevent injury?

- Some people do not like wearing seat belts because:
  - there is a risk of chest injury
  - they may be trapped in a fire
  - drivers may be encouraged to drive less carefully because they know they have the protection from a seat belt.

- All safety features must be kept in good repair.
  - Seat belts must be replaced after a crash in case the fabric has been overstretched.
  - The safety cage must be examined for possible damage after a crash.
  - Seat fixings should be checked frequently to make sure they are secure.

- Despite **computer modelling**, crash tests using real vehicles and **dummies** provide more safety information.

Improve your grade

Safety features

Explain how crumple zones help to reduce injuries in a crash. *AO1* [3 marks]

Falling safely

Falling objects

G–E

- A falling object:
  - gets faster as it falls
  - is pulled towards the centre of Earth by a force called **weight** which is caused by the **force** of **gravity**.
- Objects that have a large area of cross-section fall more slowly. For example, if a ball and a feather are dropped, the ball hits the ground first. This slowing down force is called **air resistance** or **drag**.

D–C

- All objects fall with the same **acceleration** due to gravity as long as the effect of air resistance is very small.
- The size of the air resistance force on a falling object depends on:
  - its **cross-sectional area** – the larger the area the greater the air resistance
  - its speed – the faster it falls the greater the air resistance.
- Air resistance has a significant effect on motion only when it is large compared to the weight of the falling object.
- The **speed** of a **free-fall** parachutist changes as he falls to Earth.

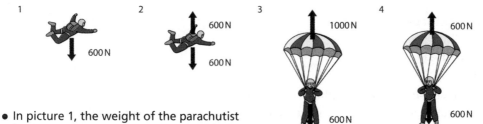

- In picture 1, the weight of the parachutist is greater than air resistance. He accelerates.
- In picture 2, the weight of the parachutist and air resistance are equal. The parachutist has reached **terminal speed** because the forces acting on him are balanced.
- In picture 3, the air resistance is larger than the weight of the parachutist. He slows down and air resistance decreases.
- In picture 4, the air resistance and weight of the parachutist are the same. He reaches a new, slower terminal speed.

G–E

- Free-falling objects:
  - don't experience air resistance
  - accelerate downwards at the same rate irrespective of **mass** or shape.
- Examples of free-falling objects are:
  - objects falling above the Earth's atmosphere
  - objects falling on the Moon
  - the Moon itself.

Friction

G–E

- All **friction** forces slow an object down.
- Frictional forces between surfaces can be reduced by lubricating the surfaces.
- Friction forces on vehicles can be reduced by streamlining their shapes by:
  - shaping car roof boxes
  - making cars wedge-like in shape
  - angling lorry deflectors
  - closing car windows when driving.

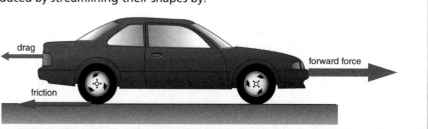

Improve your grade

Streamlining

Explain the effect a streamlined shape has on the motion of a car? *AO1* [2 marks]

The energy of games and theme rides

Gravitational potential energy

- An object held above the ground has **gravitational potential energy**.
- The amount of gravitational potential energy an object has depends on:
  - its **mass**
  - its height above the ground.

G–E

- This is shown by the formula:
  GPE = mgh **where**
  m = mass, h = vertical height moved, g = **gravitational field strength** (10 N/kg)
- GPE is measured in **joules** (J).

D–C

Energy transfers

- A bouncing ball converts gravitational potential energy to **kinetic energy** and back to gravitational potential energy. It does not return to its original height because energy is transferred.

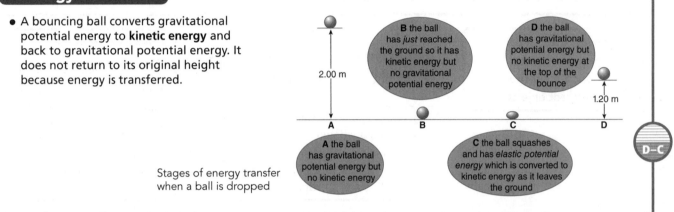

Stages of energy transfer when a ball is dropped

B the ball has *just* reached the ground so it has kinetic energy but no gravitational potential energy

D the ball has gravitational potential energy but no kinetic energy at the top of the bounce

A the ball has gravitational potential energy but no kinetic energy

C the ball squashes and has *elastic potential energy* which is converted to kinetic energy as it leaves the ground

2.00 m

1.20 m

D–C

- A water-powered *funicular railway* has a carriage that takes on water at the top of the hill giving it extra gravitational potential energy.
- As the carriage travels down the hillside it transfers gravitational potential energy to kinetic energy.
  - At the same time, it pulls up another carriage with an empty water tank on a parallel rail.

How a roller coaster works

- A roller coaster uses a motor to haul a train up in the air. The riders at the top of a roller coaster ride have a lot of gravitational potential energy.
- When the train is released it converts gravitational potential energy to kinetic energy as it falls. This is shown by the formula:

  loss of gravitational potential energy (GPE) = gain in kinetic energy (KE)

- Each peak is lower than the one before because some energy is transferred to heat and sound due to **friction** and **air resistance**.

D–C

- This is shown by the formula:

  GPE at top = KE at bottom + energy transferred (to heat and sound) due to friction

- If **speed** doubles, KE quadruples (KE $\propto v^2$).
- If mass doubles, KE doubles (KE $\propto m$).

Improve your grade

Energy changes on a water chute

Janie is using the water chute at the swimming pool. The fall of the chute is 12 m.
Janie's mass is 50 kg. The gravitational field strength is 10 N/kg.

(a) What will be her kinetic energy at the bottom of the chute as she enters the pool?

(b) Her speed as she enters the pool is twice her speed when she is 9 m above the pool.
What is her kinetic energy when she is 9 m above the pool?
AO1 [2 marks], *AO2* [2 marks]

P3 Summary

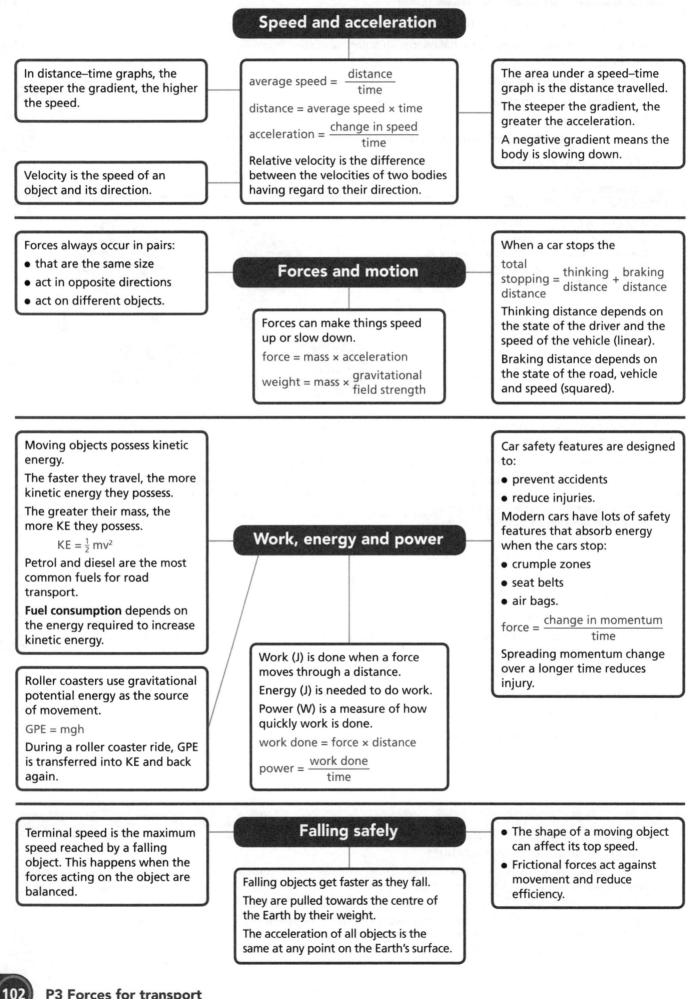

Speed and acceleration

In distance–time graphs, the steeper the gradient, the higher the speed.

Velocity is the speed of an object and its direction.

$$\text{average speed} = \frac{\text{distance}}{\text{time}}$$

distance = average speed × time

$$\text{acceleration} = \frac{\text{change in speed}}{\text{time}}$$

Relative velocity is the difference between the velocities of two bodies having regard to their direction.

The area under a speed–time graph is the distance travelled.

The steeper the gradient, the greater the acceleration.

A negative gradient means the body is slowing down.

Forces and motion

Forces always occur in pairs:
- that are the same size
- act in opposite directions
- act on different objects.

Forces can make things speed up or slow down.

force = mass × acceleration

$$\text{weight} = \text{mass} \times \text{gravitational field strength}$$

When a car stops the

$$\text{total stopping distance} = \text{thinking distance} + \text{braking distance}$$

Thinking distance depends on the state of the driver and the speed of the vehicle (linear).

Braking distance depends on the state of the road, vehicle and speed (squared).

Work, energy and power

Moving objects possess kinetic energy.

The faster they travel, the more kinetic energy they possess.

The greater their mass, the more KE they possess.

$$KE = \tfrac{1}{2}mv^2$$

Petrol and diesel are the most common fuels for road transport.

Fuel consumption depends on the energy required to increase kinetic energy.

Roller coasters use gravitational potential energy as the source of movement.

GPE = mgh

During a roller coaster ride, GPE is transferred into KE and back again.

Work (J) is done when a force moves through a distance.

Energy (J) is needed to do work.

Power (W) is a measure of how quickly work is done.

work done = force × distance

$$\text{power} = \frac{\text{work done}}{\text{time}}$$

Car safety features are designed to:
- prevent accidents
- reduce injuries.

Modern cars have lots of safety features that absorb energy when the cars stop:
- crumple zones
- seat belts
- air bags.

$$\text{force} = \frac{\text{change in momentum}}{\text{time}}$$

Spreading momentum change over a longer time reduces injury.

Falling safely

Terminal speed is the maximum speed reached by a falling object. This happens when the forces acting on the object are balanced.

Falling objects get faster as they fall.

They are pulled towards the centre of the Earth by their weight.

The acceleration of all objects is the same at any point on the Earth's surface.

- The shape of a moving object can affect its top speed.
- Frictional forces act against movement and reduce efficiency.

Sparks

Electrons

- Metals are good **electrical conductors**. They allow electric **charges** to move through them.
- Materials such as wood, glass and polythene are **insulators**. They do not allow electric charges to pass through them.
- Charge can build up on an insulator. An insulator can be charged by **friction**.
- There are two kinds of electric charge, positive and negative. When rubbed with a duster:
  - acetate and perspex become positively charged
  - polythene becomes negatively charged.

G–E

- An **atom** consists of a small positively charged **nucleus** surrounded by an equal number of negatively charged **electrons**.
- In a stable, neutral atom there are the same amounts of positive and negative charges.
- All electrostatic effects are due to the movement of electrons.
- The law of electric charge states that: like charges **repel**, unlike charges **attract**.

D–C

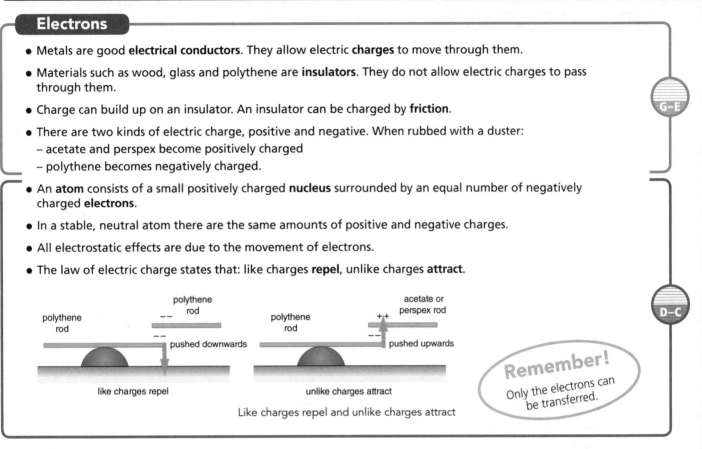

Like charges repel and unlike charges attract

Remember!
Only the electrons can be transferred.

Electric shocks

- A person gets an electric **shock** if they become charged and then become **earthed**.
- A person can become charged if they walk on a nylon-carpeted or vinyl floor because:
  - the floor is an insulator
  - the person becomes charged as they walk, due to *friction*.
- The person can become earthed by touching water pipes or even another person.
- Synthetic clothing can also cause a person to become charged.

G–E

- When inflammable gases or vapours are present, or there is a high concentration of oxygen, a **spark** from *static electricity* could ignite the gases or vapours and cause an **explosion**.
- If a person touches something at a high **voltage**, large amounts of electric charge may flow through their body to earth.
- Even small amounts of charge flowing through the body can be fatal.
- Static electricity can be a nuisance but not dangerous.
  - Dust and dirt are attracted to insulators, such as television screens.
  - Clothes made from synthetic materials often 'cling' to each other and to the body.

D–C

Improve your grade

Static charge
Connor is in the library walking on a nylon carpet. He touches a metal bookshelf and receives an electric shock. Explain how he became charged and why he received a shock.
AO2 [3 marks]

Uses of electrostatics

Uses of static electricity

G–E

- A paint sprayer **charges** paint droplets to give an even coverage.
- Crop sprayers on farms work by charging the fertiliser or insecticide in a similar way.
- A photocopier and **laser** printer use charged particles to produce an image.
- Charged plates inside factory chimneys are used to remove dust particles from smoke.
- A **defibrillator** delivers a controlled electric **shock** through a patient's chest to restart their heart.

Dust precipitators

D–C

- A dust precipitator removes harmful particles from the chimneys of factories and power stations that **pollute** the **atmosphere**.
- A metal grid (or wires) is placed in the chimney and given a large charge from a high-**voltage** supply.
- Plates inside the chimney are **earthed** and gain the opposite charge to the grid.
- As the dust particles pass close to the grid, they become charged with the same charge as the grid.
- Like charges **repel**, so the dust particles are repelled away from the wires. They are **attracted** to the oppositely charged plates and stick to them.
- At intervals the plates are vibrated and the dust falls down to a collector.

Paint spraying

D–C

- Static electricity is used in paint spraying.
  - The spray gun is charged.
  - All the paint particles become charged with the same charge.
  - Like charges repel, so the paint particles spread out giving a fine spray.
  - The object to be painted is given the opposite charge to the paint.
  - Opposite charges attract, so the paint is attracted to the object and sticks to it.
  - The object gets an even coat, with limited paint wasted.

nozzle is charged up positively

object to be painted is negatively charged

Electrostatic principles in paint spraying

Defibrillators

D–C

- Two **paddles** are charged from a high-voltage supply.
- They are then placed firmly on the patient's chest to ensure good electrical contact.
- Electric charge is passed through the patient to make their heart contract.
- Great care is taken to ensure that the operator does not receive an electric shock.

Improve your grade

Defibrillators

Explain how a human heart can be restarted using static electricity. *AO1* [3 marks]

P4 Radiation for life

Safe electricals

Resistance

- A closed loop, with no gaps, is required for a circuit to work. Charge cannot flow across a gap in a circuit.

- A **resistor** is added to a circuit to change the amount of **current** in it.

- **Resistance** is measured in **ohms** (Ω).

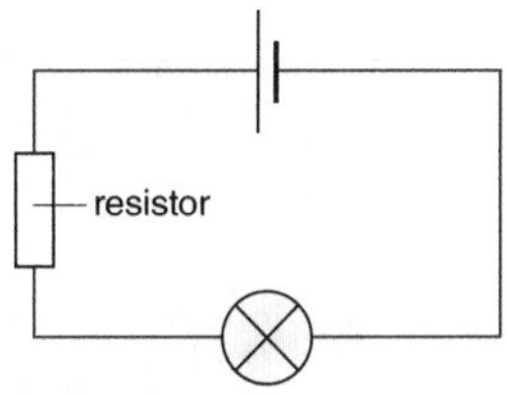

- A **variable resistor,** or **rheostat**, changes the **resistance**. Longer lengths of wire have more resistance; thinner wires have more resistance.

- **Voltage** (**potential difference**) is measured in **volts** (V) using a **voltmeter** connected in *parallel*.
  - For a fixed resistor, as the voltage across it increases, the current increases.
  - For a fixed **power** supply, as the resistance increases, the current decreases.

- The formula for resistance is:

$$\text{resistance} = \frac{\text{voltage}}{\text{current}} \qquad R = \frac{V}{I}$$

Adding a resistor to a circuit decreases the current and the lamp is dim

Remember!
Always remember to include the correct unit with your answer.

Remember!
Longer wires mean less current will pass.
Thinner wires mean less current will pass.

G–E

D–C

Live, neutral and earth wires

- The cable used to connect an appliance to the mains has three wires inside it. **Live**, **neutral** and **earth wires** can be seen in a plug. An earthed conductor can't become live.

- The colours of the wires in the three-pin plug are:
  - live – brown
  - neutral – blue
  - earth – green and yellow striped.

- **Circuit breakers** will turn off a circuit in the mains if there is a fault. They can be reset, by means of a switch, when the fault has been repaired.

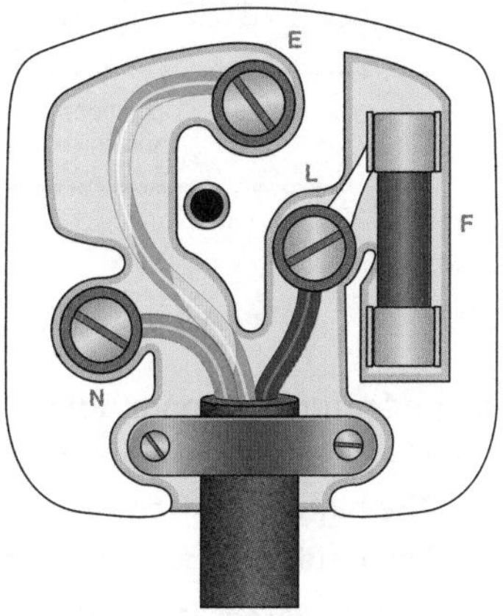

- The live wire carries a high voltage around the house.

- The neutral wire completes the circuit, providing a return path for the current.

- The earth wire is connected to the case of an appliance to prevent it becoming live.

- A **fuse** contains wire which melts, breaking the circuit, if the current becomes too large.

- No current can flow, preventing overheating and further damage to the appliance.

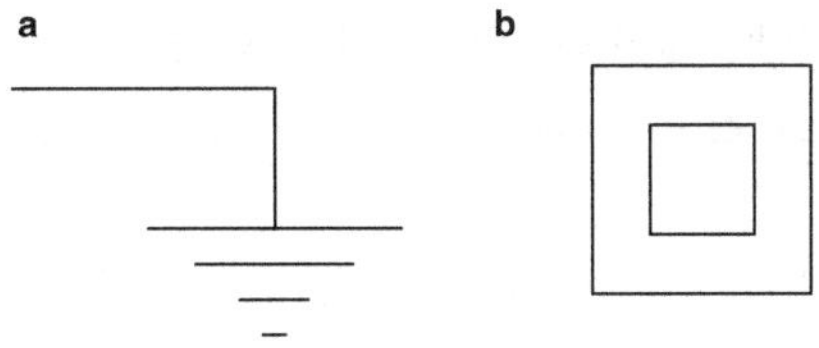

Wiring inside a three-pin plug

a   **b**

a Earth symbol and **b** double-insulation symbol.

Remember!
Double insulated appliances do not need an earth wire as the outer case is not a conductor.

G–E

D–C

Electrical power

- The rate at which an appliance transfers energy is its *power* rating:

  power = voltage × current

D–C

Improve your grade

Variable resistors

Explain how a variable resistor works and give an example of where one is used in the home. *AO1* [2 marks]

Ultrasound

Longitudinal waves

- Sound and **ultrasound** are **longitudinal waves**.

- **Wavelength** (λ) is the distance occupied by one complete wave.

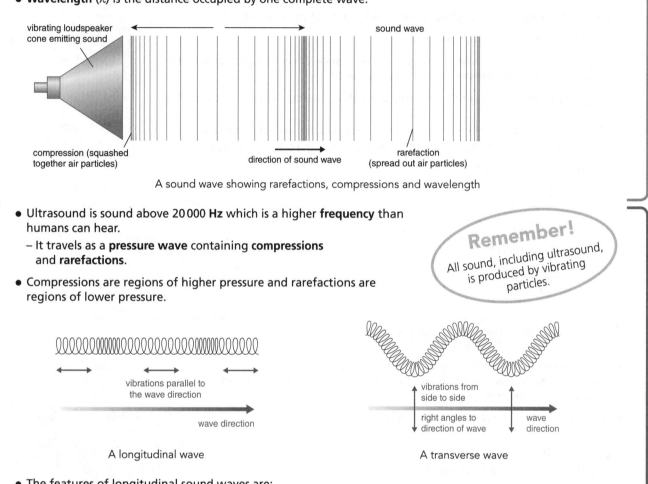

A sound wave showing rarefactions, compressions and wavelength

- Ultrasound is sound above 20 000 **Hz** which is a higher **frequency** than humans can hear.
  - It travels as a **pressure wave** containing **compressions** and **rarefactions**.

- Compressions are regions of higher pressure and rarefactions are regions of lower pressure.

> **Remember!**
> All sound, including ultrasound, is produced by vibrating particles.

A longitudinal wave

A transverse wave

- The features of longitudinal sound waves are:
  - They can't travel through a **vacuum**. The denser the medium, the faster a wave travels.
  - The higher the frequency or **pitch**, the smaller the *wavelength*.
  - The louder the sound, or the more powerful the ultrasound, the more energy is carried by the wave and the larger its **amplitude**.

Uses of ultrasound

- Ultrasound allows a doctor to 'see' inside a patient without surgery.

- Uses of ultrasound for *diagnostic* purposes include to:
  - check the condition of a foetus
  - investigate heart and liver problems
  - look for **tumours** in the body
  - measure the **speed** of blood flow in vessels when a blockage of a vein or artery is suspected.

- When ultrasound is used to break down kidney stones:
  - a high-powered ultrasound beam is directed at the kidney stones
  - the ultrasound energy breaks the stones down into smaller pieces
  - the tiny pieces are then excreted from the body in the normal way.

Improve your grade

Uses of ultrasound
Ultrasound is used to treat kidney stones.
(a) Give two other medical uses of ultrasound.
(b) Explain how kidney stones are treated. *AO1* [4 marks]

What is radioactivity?

Radioactive decay

- **Radiation** comes from the **nucleus** of **atoms**.
- A *Geiger-Muller tube* and ratemeter (together commonly called a *Geiger counter*) are used to detect the rate of decay of a **radioactive** substance.
  - Each 'click' or extra number on the display screen represents the decay of one nucleus.
  - A decaying nucleus emits radiation.
- The activity is measured by the average number of nuclei that decay every second.
  - This is also called the **count rate**.
  - Activity is measured in counts/second or becquerels (Bq).
- The formula for activity is:

$$\text{activity} = \frac{\text{number of nuclei which decay}}{\text{time taken in seconds}}$$

- The activity of a radioactive substance decreases with time – the count rate falls.
- Nuclear radiation **ionises** materials and this can damage living cells.

G–E

- Radioactive substances decay naturally, giving out **alpha**, **beta** and **gamma** radiation.
- Nuclear radiation causes **ionisation** by removing **electrons** from atoms or causing them to gain electrons.
- Radioactive decay is a **random** process; it isn't possible to predict exactly when a nucleus will decay.
- There are so many atoms in even the smallest amount of radioisotope that the average count rate will always be about the same. Radioisotopes have unstable nuclei. Their nuclear particles aren't held together strongly enough.
- The **half-life** of a radioisotope is the average time for half the nuclei present to decay. The half-life cannot be changed.

Remember!
Alpha, beta and gamma radiation are not radioactive; it's the source that emits them that is radioactive.

D–C

What are alpha and beta particles?

- When an alpha or a beta particle is emitted from the nucleus of an atom, the remaining nucleus is a different element.

D–C

- Alpha particles are very good ionisers. They are the largest particles emitted in radioactive decay. This means they are more likely to strike atoms of the material they are passing through, ionising them.

G–E

	alpha (α) particle	beta (β) particle
properties	positively charged	negatively charged
	has a large **mass**	has a very small mass
	is a **helium** nucleus	travels very fast
	has helium gas around it	is an electron
	consists of two **protons** and two **neutrons**	

D–C

- When an alpha or a beta particle is emitted from the nucleus of an atom, the remaining nucleus is a different element. But gamma radiation it is not a particle, so does not change the composition of the nucleus; it remains the same element.

EXAM TIP
You must know the difference between alpha, beta and gamma radiation.

Improve your grade

Nuclear radiation
What is an alpha particle? *AO1* [1 mark]

Uses of radioisotopes

Background radiation

G–E
- **Background radiation** is **ionising** radiation that is always present in the environment.
  - It varies from place to place and from day to day.
  - The level of background radiation is low and doesn't cause harm.

- Background radiation is due to:

D–C
  - **radioactive** substances present in rocks (especially **granite**) and soil
  - **cosmic rays** from Space
  - man-made sources including **radioactive waste** from industry and hospitals.

Remember!
Always subtract background radiation from any measurement of activity of a radioactive source.

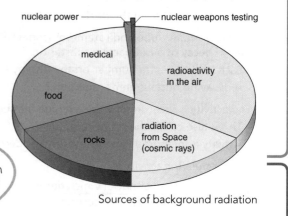

Sources of background radiation

Tracers

G–E
- Radioisotopes are used as **tracers** in industry and in hospitals (see page 109),
  - to track dispersal of waste
  - to find leaks/blockages in underground pipes
  - to find the route of underground pipes.

D–C
- When using a tracer to locate a leak in an underground pipe:
  - a very small amount of a **gamma** emitter is put into the pipe
  - a detector is passed along the ground above the path of the pipe
  - an increase in activity is detected in the region of the leak and little or no activity is detected after this point.

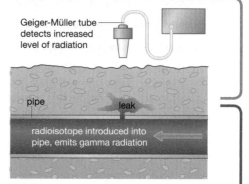

Using a radioisotope to detect a leak in a pipe

Smoke detectors

G–E
- One type of **smoke detector** uses a source of alpha particles to detect smoke. It's sensitive to low levels of smoke.

D–C
- A smoke detector contains an **isotope** which emits **alpha** particles.
  - Without smoke, the alpha particles **ionise** the air which creates a tiny **current** that can be detected by the circuit in the smoke alarm.
  - With smoke, the alpha particles are partially blocked so there is less ionisation of the air. The resulting change in current is detected and the alarm sounds.

Dating rocks

G–E
- Radioactivity can be used to date rocks.

D–C
- Some rock types such as granite contain traces of **uranium**, a radioactive material.
  - The uranium isotopes present in the rocks go through a series of decays, eventually forming a stable isotope of **lead**.
  - By comparing the amounts of uranium and lead present in a rock sample, its approximate age can be found.

Radiocarbon dating

D–C
- **Carbon-14** is a radioactive isotope of carbon that is present in all living things. By measuring the amount of carbon-14 present in an archaeological find, its approximate age can be found.

Improve your grade

Tracers
How are radioactive sources used to find blockages in underground pipes? *AO1* [3 marks]

Treatment

Using radiation

- **X-rays** and **gamma** (γ) **rays** are used in medicine:
  - for *diagnosis* and for **therapy**.
- Gamma **radiation** is emitted from the **nucleus** of an **atom**.
- Medical radioisotopes are produced by placing materials into a nuclear **reactor**.

- Radiation emitted from the nucleus of an unstable atom can be **alpha** (α), **beta** (β) or gamma (γ).
  - Alpha radiation is absorbed by the skin so is of no use for diagnosis or therapy.
  - Beta radiation passes through skin, but not bone. Its medical applications are limited but it is used, for example, to treat the eyes.
  - Gamma radiation is very penetrating and is used in medicine. *Cobalt-60* is a gamma-emitting **radioactive** material that is widely used to treat **cancers**.
- When nuclear radiation passes through a material it causes **ionisation**. Ionising radiation damages living cells, increasing the risk of cancer.
- Cancer cells within the body can be destroyed by exposing the affected area to large amounts of radiation. This is called **radiotherapy**.
- Materials can be made radioactive when their nuclei absorb extra **neutrons** in a nuclear reactor.

Gamma radiation

- Gamma radiation is used to treat cancer.
  - Large doses of high-energy radiation can be used in place of surgery. However, it's more common to use radiation after surgery, to make sure all the cancerous cells are removed or destroyed.
  - If any cancerous cells are left behind, they can multiply and cause secondary cancers at different sites in the body.
- Gamma radiation kills bacteria. It's used to sterilise hospital equipment to prevent the spread of disease.

Comparing x-rays and gamma rays

- Gamma rays and x-rays are both ionising electromagnetic waves.
- Gamma rays and x-rays have similar **wavelengths** but are produced in different ways.
- When x-rays pass through the body the tissues absorb some of this ionising radiation. The amount absorbed depends on the thickness and the **density** of the absorbing material.

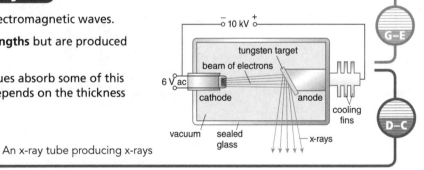

An x-ray tube producing x-rays

Tracers

- A radioactive **tracer** is used to investigate inside a patient's body without surgery.
  - Technetium-99m is a commonly used tracer. It emits only gamma radiation.
  - **Iodine**-123 emits gamma radiation. It is used as a tracer to investigate the **thyroid gland**.
  - The radioactive tracer being used is mixed with food or drink or injected into the body.
  - Its progress through the body is monitored using a detector such as a gamma camera connected to a computer.

Radiographer

- A *radiographer* is a person who carries out procedures using x-rays and nuclear radiation. They must ensure that they are not exposed to x-rays or nuclear radiation. They can do this by leaving the room while the radiation source is switched on.

Improve your grade

Medical tracers

Why is iodine-123 used as a tracer in medicine? *AO1* [3 marks]

Nuclear power stations

G–E

- A **power station** produces electricity.
- It needs an **energy** source such as coal, oil, gas or nuclear:
  - to heat water
  - that produces steam
  - that turns a **turbine**
  - to generate electricity.
- A **nuclear power station** uses **uranium** as a fuel instead of burning coal, oil or gas to heat the water.
- The process that gives out energy in a nuclear reactor is **fission**, and this must be kept under control.
- Nuclear fission produces **radioactive waste** which has to be handled carefully and disposed of safely.

D–C

- Natural uranium consists of two **isotopes**, uranium-235 and uranium-238.
  - The 'enriched uranium' used as fuel in a nuclear power station contains a greater proportion of the uranium-235 isotope than occurs naturally.
- Fission occurs when a large **unstable nucleus** is split up and energy is released as heat.
  - The heat is used to boil water to produce steam.
  - The pressure of the steam acting on the turbine blades makes it turn.
  - The rotating turbine turns the **generator**, producing electricity.
- When uranium fissions, a **chain reaction** starts. A nuclear bomb is an example of a chain reaction that is not controlled.

Fusion

G–E

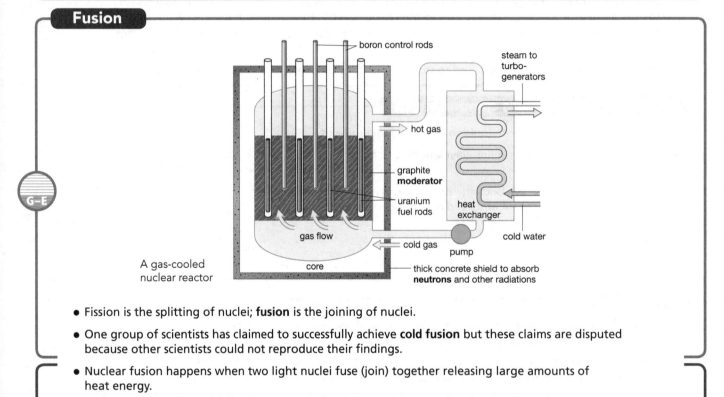

A gas-cooled nuclear reactor

- Fission is the splitting of nuclei; **fusion** is the joining of nuclei.
- One group of scientists has claimed to successfully achieve **cold fusion** but these claims are disputed because other scientists could not reproduce their findings.

D–C

- Nuclear fusion happens when two light nuclei fuse (join) together releasing large amounts of heat energy.
- Fusion requires extremely high **temperatures** which have proved difficult to achieve and to manage safely on Earth and so its use for large-scale power generation remains a dream.
- Research in this area is very expensive so it is carried out as an international joint venture to share costs, expertise and benefits.

Improve your grade

Nuclear fission

Explain how nuclear fission produces electricity. *AO1* [3 marks]

P4 Summary

Electrostatics

Static electricity can be dangerous in some situations.

Electrostatic effects are caused by the transfer of electrons.

Uses of electrostatics include:
- defibrillators
- paint and crop sprayers
- dust precipitators
- photocopiers.

Metals are good conductors so charges do not build up on them.

Insulators can be charged by friction.

There are two kinds of electric charge, positive and negative.

Like charges repel, unlike charges attract.

Using electricity safely

In a three-pin plug:
- live wire is brown – it is at a high voltage
- neutral wire is blue – it completes the circuit
- earth wire is green/yellow – it is connected to the case to prevent it becoming live.

Double insulated appliances do not need an earth wire.

power = voltage × current

The fuse is in the live wire. If the current is greater than the rating, the fuse will melt (or circuit breaker switch) breaking the circuit.

$$resistance = \frac{voltage}{current}$$

Longer wires have more resistance.

Thinner wires have more resistance.

Ultrasound

Medical uses include:
- scans to see inside the body without surgery
- measuring the rate of blood flow
- breaking up kidney stones.

Ultrasound is sound above 20 000 Hz which is a higher frequency than humans can hear.

Longitudinal waves – e.g. sound and ultrasound.

Transverse waves – e.g. light.

Wavelength (λ) is the distance occupied by one complete wave.

Frequency (f), measured in hertz (Hz), is the number of complete waves in 1 second.

Nuclear radiation

Radioactivity comes from the nucleus of atoms.

Background radiation is always in the environment, from rocks, soil and cosmic rays.

Activity is the average number of nuclei decaying in one second.

Fission is the splitting up of a large nucleus.

Fusion is the joining together of smaller nuclei.

Both release a lot of energy.

Fission is currently used to generate electricity.

Fission leads to a chain reaction which must be carefully controlled.

Nuclear radiation is emitted from the nuclei of radioisotopes:
- α particle is a helium nucleus
- β particle is a fast-moving electron
- γ radiation is a short wavelength electromagnetic wave.

The half-life is the average time for half of the nuclei present to decay.

Medical uses of radioactivity include:
- diagnosis, as a tracer
- sterilising equipment
- treating cancers.

Only β and γ can pass through skin. γ radiation is most widely used as it is the most penetrating.

Other uses include:
- smoke detectors
- industrial tracers
- dating rocks and archaeological finds.

B1 Improve your grade Understanding organisms

Page 4 Healthy and fit?
Joe thinks he is healthy but not fit. Explain the difference.
AO1 [3 marks]

Joe may be unfit to play football but not have a cold.

Answer grade: G–F. The answer hints at the differences between being healthy and being fit. For full marks, describe what being healthy means, e.g. having no diseases such as those caused by bacteria and viruses. Also, explain that fitness can be measured in many ways, e.g. by measuring strength, flexibility or speed.

Page 5 Balanced diet
Charlie is fourteen and does athletics. She eats the same diet as her mother. Her mother's diet is a balanced one but Charlie's is not. Explain why. *AO1/2* [4 marks]

A balanced diet contains many things. Charlie is much younger than her mother and needs different things.

Answer grade: G–F. For full marks, also state that a balanced diet contains carbohydrates, fats, proteins, minerals, vitamins, fibre and water. Point out that Charlie is an active teenager, so she will need more carbohydrates and fats for energy and protein for growth.

Page 6 Drug trials
Some scientists believe they have discovered a new drug to treat malaria. Suggest why it may take up to ten years before the drug is available for general use. *AO2* [4 marks]

The scientists would have to carry out human trials over many years to be sure that the drug worked.

Answer grade: F–E. For full marks, explain that tests using animals, human tissue and computer models are done before human trials. Trials take time because they must ensure safety and no long-term effects; and set up mass production of the drug once is it tested.

Page 7 Monocular and binocular vision
Predators such as owls hunt and catch their prey. They have good binocular vision. Explain the advantages and disadvantages of binocular vision compared with monocular vision. *AO1* [4 marks]

Binocular vision means that owls are able to judge distance very well to catch their prey. However, they have a narrow field of view.

Answer grade: G–E. This gives only one advantage and one disadvantage of binocular vision. For full marks, compare monocular and binocular vision. State that in binocular vision the brain compares images from each eye: the more similar the images, the further away they are.

Page 8 Smoking
Asif smokes cigarettes. Suggest why he finds it difficult to give up smoking and list the extra risks of being a smoker. *AO1* [4 marks]

Asif is used to smoking and so finds it difficult to give up. He will be at greater risk of getting lung cancer.

Answer grade: G. This answer is too vague. For full marks, state that nicotine in cigarette smoke is addictive and the absence of nicotine causes withdrawal symptoms, making it difficult to stop. Include a full list of conditions with extra risk: emphysema, bronchitis, heart disease; and cancer of the mouth, throat, oesophagus and lung.

Page 9 Diabetes
Describe and explain the link between the pancreas and Type 1 diabetes. *AO1/2* [4 marks]

The pancreas produces insulin. Type 1 diabetes is linked to insulin.

Answer grade: G. For full marks, explain that insulin is a hormone that is produced by the pancreas and carried round the body in the blood. Also state that Type 1 diabetes is caused by the pancreas not producing insulin, which means that sugar levels in the blood cannot be controlled.

Page 10 Phototropism
Look at the diagram (on page 10) showing an experiment on cress seedlings. What is the experiment designed to find out? Explain the plants' reactions. *AO2* [4 marks]

The experiment is trying to find out how cress seedlings react to light. The seedlings in the box grew towards the light opening. The seedlings on the clinostat grew straight upwards.

Answer grade: G–E. The results are described rather than explained. For full marks, explain that the growth of the shoots towards light shows that they are positively phototropic and that this reaction is caused by chemicals called plant hormones.

Page 11 Breeding tomato plants
Look at the diagram (on page 11) of a breeding experiment using tomato plants.
(a) Where is the genetic material about stem colour held in the tomato plants?
(b) Explain the results of the F1 generation. *AO1/2* [5 marks]

(a) The genetic information is found in genes.
(b) The F1 generation are all purple stemmed.

Answer grade: F–E. In (a), the results of the F1 generation are described but not explained. For full marks, state that the chromosomes in the cell's nucleus carry the information in the form of genes. In (b), state that the F1 generation are all purple stemmed though they all carry the recessive characteristic of being green stemmed; the purple characteristic is dominant over the recessive green colour.

Page 13 Classifying newly discovered organisms

A new group of animals is discovered. What steps would scientists take to work out if it is a new species and to give it a name? AO1 [4 marks]

Scientists would look at the characteristics of the animals. They would see if they are different to other animals and then give it a Latin name.

Answer grade: G–F. The answer states importance of looking at characteristics and that a binomial name is in Latin. However, there is no mention of the ability of a species to interbreed. For full marks, state that members of the same species can breed to produce fertile offspring. Include a description of the binomial system.

Page 14 Pyramids of biomass

It is sometimes hard to work out which trophic level to place an organism in when drawing a pyramid of numbers. Even if it is possible, it does not always form a pyramid. Explain these observations. AO2 [4 marks]

Some animals eat lots of different types of food. It is easy to make mistakes when drawing a pyramid.

Answer grade: G–F. For full marks, state that as some animals eat a variety of food they may be feeding at more than one trophic level. Pyramids of numbers are not always pyramids because they do not take into account the size of the organism.

Page 15 The nitrogen cycle

Explain how nitrogen in a protein molecule in a dead leaf can become available again to a plant. AO1 [4 marks]

The dead leaves rot away in the soil. This releases minerals, which can be taken up by the roots of the plants.

Answer grade: G–F. The answer states that the plant leaves have to rot to release minerals but it does not say which minerals contain nitrogen or what rots the leaves. For full marks, also state that decomposers break down proteins and that plants take up nitrates.

Page 16 Fish and shrimp

On coral reefs there are areas where shrimps live. Fish regularly visit these areas and allow the shrimps to move over their bodies, feeding. What type of relationship is this and what is the benefit to each organism? AO2 [2 marks]

This is a relationship where both gain. The shrimp gets food and the fish get cleaned.

Answer grade: G–F. The answer does not give the correct term for this relationship and the use of the word 'cleaned' is too vague. For full marks, the answer must say that this is mutualism and that the fish has its parasites removed by the shrimp.

Page 17 Living in hot, dry conditions

A camel has wide feet, skin with few hairs underneath the body and the ability to produce concentrated urine. Suggest how these features are an advantage to living in hot, dry areas. AO2 [3 marks]

The wide feet will help the camel move. The lack of hair will keep the camel cool. Producing a concentrated urine will mean it does not have to drink much.

Answer grade: G–F. The explanations are incomplete. For full marks, discuss how the wide feet will spread the weight of the camel and stop it sinking in the sand. The lack of hair will stop the animal overheating, as it will allow heat loss. Producing a concentrated urine will help conserve water, which is in short supply.

Page 18 Explanations for evolution

Human ancestors had more hair than modern humans. One idea why modern humans have less hair is that it might make them less likely to have parasites. Use Darwin's ideas to explain why modern humans have less hair than their ancestors. AO2 [4 marks]

Humans with less hair are less likely to get parasites and so this is an advantage.

Answer grade: G–F. The answer states that having less hair could be an advantage but not how it may have been caused or why it became more common. For full marks, explain that mutations produce genes for less hair and that these genes are more likely to be passed on.

Page 19 Population and pollution

Explain the reasons for the increase in carbon dioxide levels in the atmosphere and explain why people concerned about this. AO1 [4 marks]

More carbon dioxide is released from burning coal. People are worried because this causes pollution.

Answer grade: G–F. The answer gives a reason for the increase of carbon dioxide and a possible consequence but neither is fully explained. For full marks, say why fossil fuels are being burned and explain some of the possible consequences of global warming.

Page 20 Saving endangered species

The Hawaiian goose is only found on the islands of Hawaii. In the mid-1900s only about 30 were left alive. The space on the islands is restricted and a number of animals have been introduced to the islands. Write about the problems facing scientists trying to save the goose from extinction AO2 [4 marks]

There is little room for them to live on the islands. The scientists might want to conserve them for a food supply.

Answer grade: G–F. The answer highlights the lack of habitat and one reason for conservation. For full marks, include reference to the number of geese falling below a critical number and give some other reasons for conservation.

Page 22 Fractional distillation

What can crude oil be separated into and how does fractional distillation work? *AO1* [4 marks]

Crude oil can be separated into fractions such as petrol, diesel and paraffin, because fractions have different boiling points.

Answer grade: G–E. These two statements are correct but for full marks the student should include ideas that the fractions with the lowest boiling points 'exit' the tower at the top, and fractions with the highest boiling points exit at the bottom. There is a temperature gradient up the tower from high temperature at the bottom to lower temperature at the top.

Page 23 Choosing a fuel

Use the table at the top of the page (page 23) to decide which fuel should be used in a car engine. Justify your answer from the evidence. *AO3* [2 marks]

Although both have good availability, we have a system of garages around the country that sell petrol which is easier to obtain than coal.

Answer grade: None. No marks would be awarded for this answer. No prior knowledge is needed to gain full marks, as the evidence is provided in the table:

* petrol flows easily in engines, which makes it easy to use
* petrol reacts instantly so would be ideal to power a car which needs to be able to respond quickly.

Other questions may ask you to use information supplied to choose the key factors needed from a fuel when used for a particular purpose.

Page 24 Evolution of the atmosphere

Describe what gases are present in the atmosphere and how the levels of these gases are maintained. *AO1* [5 marks]

The main gases of the atmosphere are nitrogen, oxygen and carbon dioxide. Plants use carbon dioxide for photosynthesis and produce oxygen. The oxygen is used up in respiration and carbon dioxide is given out.

Answer grade: G–E. Many good points are included here. However, for full marks in the answer the student should have included the percentages of the main gases of the atmosphere (78% nitrogen, 21% oxygen and 0.035% carbon dioxide) and explained how combustion uses up oxygen but gives out carbon dioxide. They should also have stated that these three processes are part of the carbon cycle.

Page 25 Interpreting displayed formulae

The displayed formula of butene is shown. How do you know butene is a hydrocarbon and what is the name of the polymer made from butene? *AO2* [2 marks]

The molecule has a carbon and hydrogen in it so it is a hydrocarbon. The butene molecule will be made into poly(butene).

Answer grade: G–E. Both points are correct. To improve your grade you may be asked how to identify that this is an alkene molecule.

Page 26 Properties and uses of polymers

Explain which of these polymers would be a good material to make drain pipes, electrical cable covers and socks.

	A	B	C
waterproof	no	yes	yes
electrical conductor	no	no	no
flexible	yes	no	yes

AO2 [3 marks]

Drain pipes need to be waterproof and rigid so would be made from B. Electrical cable covers need to be electrical insulators and flexible and waterproof so will be made from C.

Answer grade: E. Two marks would be given for this answer. To improve your grade you will need to be able to say that the material for socks needs to be flexible but not waterproof so will be made from A.

Page 27 Using baking powder

How does baking powder, sodium hydrogencarbonate, help cakes to rise? Write a word equation for the reaction that happens. *AO1* [3 marks]

Baking powder gives off carbon dioxide when it is heated which helps cakes rise.

Answer grade: G–E. The answer is correct. For full marks, a word equation should also have been included: sodium hydrogencarbonate → sodium carbonate + carbon dioxide + water

Page 28 Esters and perfumes

Explain what specific properties a perfume needs as well as a pleasant smell. *AO1* [3 marks]

A perfume must evaporate easily, be non-toxic, not react with water, be insoluble in water and not irritate the skin.

Answer grade: G–E. All the points about perfume properties are made but for full marks the student should also have explained why they are needed, i.e. it must evaporate easily so that the perfume particles can reach the nose; be non-toxic so it does not poison you; not react with water so the perfume does not react with perspiration; be insoluble in water so it cannot be washed off easily; and not irritate the skin so the perfume can be put directly on the skin.

Page 29 Drying paint

Explain what paint is made up of and how phosphorescent pigments work. *AO1* [3 marks]

A paint is made of a binding medium, which sticks the pigment to the surface, a pigment, which gives the colour, and a solvent, which makes the paint easier to spread.

Answer grade: G–E. The points are correct but for full marks the student should also have explained that phosphorescent pigment glows in the dark.

Page 31 Living near a volcano

When a volcano erupts, it releases lava. What affects the type of volcanic eruption? *AO1* [3 marks]

If the lava is the runny kind, it will erupt slowly. If the lava is thick, the eruption will be violent and catastrophic.

Answer grade: G–E. Both sentences would gain a mark. For full marks, the answer needs to explain that the type of lava formed (and therefore the type of volcanic eruption) depends on the composition of the magma.

Page 32 Rock hardness

Describe how concrete can be made. *AO1* [2 marks]

Concrete is made by mixing together cement, sand, small stones and water, which is then allowed to set.

Answer grade: G–E. All the explanation so far is correct and worth two marks. To improve your grade you also may be asked how cement is made: cement is made from limestone and clay, which are heated together.

Page 33 Uses of metals

Three metals have properties on a relative scale where 1 is the least and 10 is the most.

Metal	Density	Strength	Hardness
aluminium	1	5	4
iron	6	8	9
lead	10	1	1

Using the information in the table, explain why you would use aluminium for aircraft bodies, iron to construct bridges and lead for protecting joins in roofs. *AO2* [3 marks]

Aluminium is less dense, iron is strong and lead is not hard so it bends to shape.

Answer grade: G–E. The matching of properties with use is correct. For full marks, the answer should provide a more detailed explanation for why they are chosen: aluminium is ideal for aircraft being lightweight and strong, and soft enough so that can be moulded; iron is dense, strong and hard for making bridge structures that carry a lot of weight; lead can be easily bent to shape for joins in roofs and very dense so that it will withstand the weather.

Page 34 Recycling cars

Suggest why people recycle materials from old cars. *AO1* [2 marks]

Recycling means materials can be used again or made into different materials. This saves natural resources so new materials do not need to be made and reduces the problems of disposing of materials.

Answer grade: G–E. This answer would gain two marks for the information on saving natural resources and reducing disposal problems. To improve your grade you may be asked about the disadvantages of recycling, such as cost and the difficulties of separating the materials.

Page 35 The manufacture of ammonia

Describe how ammonia is manufactured by the Haber process. *AO1* [4 marks]

Nitrogen reacts with hydrogen to make ammonia. The nitrogen comes from the air and the hydrogen comes from natural gas.

Answer grade: G–E. The reaction and the raw materials are correct. For full marks, the answer should also state the conditions needed: 450 °C temperature, high pressure and a catalyst, and state that the reaction is reversible.

Page 36 Estimating pH

Each of four solutions – A, B, C and D – have a different pH.

solution	A	B	C	D
pH	2	6	7	9

Explain how you can estimate whether these pH values are correct and what happens if B and D are mixed. *AO2* [3 marks]

Put universal indicator in each sample. A will go red, B will go yellow, C will go green and D will go blue. If B and D are mixed they will go a colour between green and blue.

Answer grade: G–E. The colours of universal indicator (UI) were well remembered but it would have been enough to say that it turned different colours in solutions of different pH and could be matched to a colour chart. For full marks, the answer should explain that different indicators turn different colours over pH ranges and some, such as universal indicator, are more specific.

Page 37 Essential elements in a fertiliser

What do fertilisers contain and how do they get into the plant? *AO1* [3 marks]

Fertilisers need to contain the essential elements of nitrogen, phosphorus and potassium and get into the plant through the root.

Answer grade: G–E. All three essential elements have been identified in the answer, and that they enter the plant by the roots. For full marks, the answer should explain that fertilisers must first be dissolved in water so that they can be absorbed by the plant.

Page 38 Splitting up sodium chloride

How can concentrated sodium chloride (salt) solution be split up, and what are the products of the reaction? *AO1* [2 marks]

It is split up by electrolysis to give off chlorine and hydrogen.

Answer grade: G–E. This answer would be awarded two marks. To improve your grade, you may be asked to identify that chlorine is made at the anode, hydrogen is made at the cathode and that sodium hydroxide is also made.

Page 40 Specific latent heat

Julie boils a kettle of water. Calculate the energy required to change 0.2 kg of water to steam at 100 °C. Specific latent heat = 2 260 000 J/kg. *AO2* [2 marks]

energy = 0.2 × 2 260 000 × 100 = 45 200 000 J

Answer grade: F–E. The calculation has included a temperature change. For full marks, the student should remember that latent heat changes take place at one temperature.

Page 41 Cavity wall insulation - payback time

(a) The Kelly family spend £250 on cavity wall insulation. The payback time is 2 years. What is meant by payback time? *AO1* [1 mark]
(b) Explain how cavity wall insulation reduces energy transfer. *AO1* [3 marks]

(a) The time it takes for them to pay for the insulation.

(b) Air is a good insulator so cuts down energy loss by conduction.

Answer grade: G–F. (a) This answer is typical of a student working at or around grades this level. Payback time depends on the cost of the cavity wall insulation and the annual saving on energy bills.

(b) This answer is typical of a candidate working at or around this level. The sentence is correct but not a full answer. It will score a mark but a full answer requires a statement that the air is trapped, so also reduces energy transfer by convection.

Page 42 Refraction

A ray of light is travelling from air into a glass block with an angle of incidence of 40°. Explain what happens to the ray of light after it enters the glass block. *AO1* [2 marks]

The ray of light changes direction in the glass block because it is now in a different material and the angle of refraction is less than 40°.

Answer grade: F–E. The sentence describes what happens but does not explain. For full marks, the student should refer to the fact that the speed of light decreases in the glass.

Page 43 Uses of lasers

Lasers can be used to cut through sheets of metal. A normal white light source will not do this. Explain why. *AO1* [2 marks]

Lasers produce an intense narrow beam of light.

Answer grade: F–E. The sentence is correct but not a full explanation. For full marks, the student should include the fact that laser light is monochromatic (single colour).

Page 44 Uses of mobile phones

The distance between mobile phone masts can vary between 0.8 km in the centre of a town up to 4 km in the countryside. Why are mobile phone masts much closer together in the centre of towns? *AO1* [1 mark]

The mobile signal will not travel as far between masts in a town.

Answer grade: G–F. The sentence is not correct. For full marks, the student should include the fact that the masts must be in line of sight.

Page 45 Analogue and digital signals

Describe the differences between analogue and digital signals. *AO1* [2 marks]

Analogue signals can have any value. They show exactly what is happening.

Answer grade: G–F. The sentences give information about analogue signals but not digital. For full marks, the student should include the fact that digital signals can only have two values – on and off.

Page 46 Radio communication

The picture shows a transmitter and receiver on the Earth's surface, out of line of sight.

Describe how long-wave radio signals travel from the transmitter to the receiver. *AO1* [2 marks]

Radio signals are reflected from the atmosphere.

Answer grade: F–E. The sentence mentions reflection from the atmosphere. For full marks, the student should include reference to refraction as well.

Page 47 Ozone depletion

The ozone layer in the atmosphere is getting thinner. This means more ultraviolet radiation reaches the Earth.

(a) Write down two effects on humans of increased exposure to ultraviolet radiation. *AO1* [2 marks]
(b) How can humans reduce the effects of exposure to ultraviolet radiation? *AO1* [1 mark]

(a) Cataracts
Cancer

(b) Use suncream

Answer grade F–E. In (a) two reasons are given but cancer is not a full enough answer. For full marks, the student should state skin cancer. In (b) a correct answer is given.

Page 49 Photocells

Describe and explain two situations in which photocells would be the ideal energy source. *AO2* [4 marks]

In a remote location, like in the middle of a desert or out in space, also in the middle of the sea on a ship, or far out in the countryside.

Answer grade: F–E. Places where photocells could be used have been suggested without any explanation. For full marks, the student should add explanations of why the places are suitable.

Page 50 Generators

Describe how electricity is generated. *AO1* [3 marks]

Electricity can be generated by using a coil and a magnet.

Answer grade: G–F. The answer correctly states that a magnet and a coil are needed to generate an electric current but does not mention movement. For full marks, the student should state that a coil has to be in a changing magnetic field.

Page 51 Earth's temperature

What effects can dust in the atmosphere have on the temperature of the Earth? *AO1* [2 marks]

Dust reflects heat back to the Sun causing global warming.

Answer grade: G–E. Only one effect of dust is mentioned. For full marks, the student should include the fact that dust can have two opposite effects: the smoke from factories reflects radiation back to Earth – the temperature rises as a result; and ash clouds from volcanoes reflect radiation from the Sun back into space – the temperature falls as a result.

Page 52 Cost of electricity

Calculate the cost of using a 2000 W hairdryer for 30 minutes to dry your hair if one kWh costs 10p. *AO2* [3 marks]

Power × time × cost per unit

2000 × 30 × 10 = 600 000 p or £6.00

Answer grade: F–E. The working is shown but incorrect units have been used. For full marks, the student should convert to kilowatts and hours before multiplying.

Page 53 Ionising radiation

Describe two medical uses of nuclear radiation. *AO1* [2 marks]

Alpha radiation is used in smoke detectors and beta radiation is used to monitor the thickness of paper in a paper mill.

Answer grade: G. Uses of nuclear radiation are given but these are not medical uses. For full marks, the student should give medical uses of radiations, e.g. 'Gamma radiation is used in the treatment of cancer and to sterilise medical instruments.'

Page 54 Travelling into space

Describe the problems astronauts will experience carrying out daily routines when they are aboard the International Space Station. *AO2* [3 marks]

Astronauts travelling to other planets will have to wear heavy spacesuits. They experience weightlessness and will have to exercise while strapped down. They also have problems eating, sleeping and using the toilet as everything floats.

Answer grade: G–F. The student has not realised that astronauts have not been as far as another planet yet and may lose marks for this. On the International Space Station the astronauts can wear their own clothes as the conditions will be made as close to Earth as possible. They will experience weightlessness and so the second part of the answer is correct.

Page 55 Comets

How do we see a comet? *AO1* [2 marks]

A comet is visible when it comes close to the Sun.

Answer grade: G–F. Although this answer is correct it does not describe the mechanism by which a comet becomes visible. For full marks, the student should include the fact that heat from the Sun melts the ice and that dust and rocks released from the comet then glow.

Page 56 Life of a star

List the last three stages a star like our Sun goes through when it is dying. *AO1* [3 marks]

Red giant, supernova, black hole.

Answer grade: D. This answer has mixed up the death of a large star with a medium star like our Sun. For full marks, the sequence red giant, white dwarf, black dwarf should be stated.

Page 58 Genes

Before Watson and Crick's work on DNA, many people used the word 'gene' but did not know what it actually was. Write about what a gene is and what it does. AO1 [4 marks]

A gene is made of DNA. Genes are found in all cells and they control the characteristics of an organism.

Answer grade: G–F. Sentence 1 correctly states that genes are made of DNA but does not mention chromosomes. Sentence 2 mentions control but does not explain how this happens. For full marks, include a description of a gene as being a section of a chromosome that codes for a particular protein.

Page 59 Enzymes

Biological washing powders contain enzymes such as lipase, which break down fat molecules in stains. Explain how enzymes such as lipase work and why biological washing powders should be used at the stated temperature. AO1 [3 marks], AO2 [2 marks]

Enzymes like lipase work as biological catalysts. If the temperature is too high the clothes will not be cleaned properly.

Answer grade: G–F. Sentence 1 correctly says that lipase is a catalyst but it does not describe how it works. Sentence 2 does not explain *why* the clothes are not cleaned. For full marks, include a description of the lock and key theory and an idea of optimum temperature.

Page 60 Training

Training improves the blood flow to the muscles of marathon runners. This helps them to respire aerobically rather than anaerobically when they run their races.

Explain why an improved blood supply helps them to do this and what advantage it gives them. AO1 [2 marks], AO2 [2 marks]

A better blood supply means more blood will get to the muscles. This means that anaerobic respiration will not happen and lactic acid will not be made.

Answer grade: G–F. Sentence 1 does not explain why a rich blood supply is so important. Sentence 2 links anaerobic respiration with lactic acid formation but it does not say why this is undesirable. For full marks, explain that a better blood supply will provide more oxygen and that anaerobic respiration releases less energy, and lactic acid, which causes muscular pain.

Page 61 Functions of mitosis

Human body cells can only divide by mitosis a certain number of times. Using the functions of mitosis, suggest what effect this has on the body. AO1 [2 marks], AO2 [2 marks]

The cells can only divide so many times and this is why we start to look old.

Answer grade: G–F. The answer links the lack of new cells with signs of ageing but it does not refer to the functions of mitosis. For full marks, it should state that mitosis is needed to replace damaged and worn out body cells. Without this, injuries are slower to heal, hairs are not replaced etc.

Page 62 Blood vessels and the heart

Explain these observations.
(a) If some blood vessels are cut, the blood oozes out, whereas if others are cut it will pulse out.
(b) The walls of the atria are much thinner than the ventricle walls. AO2 [4 marks]

(a) *If an artery is cut, the blood pulses out. If it is a vein, it oozes out.*
(b) *The ventricles have to pump the blood further than the atria.*

Answer grade: G–F. (a) correctly identifies the blood vessels but does not explain the difference. (b) gives a correct reason but again does not give a full explanation. For full marks in (a), the answer should include the difference in pressure between the blood in arteries and veins, and (b) would need to explain that a thicker wall means more muscle and so greater pressure can be generated.

Page 63 Plant growth

Katie wants to investigate the growth of some broad bean seeds.
(a) Describe two different measurements she could use to plot the growth of the young plants.
(b) Describe how the pattern of growth in these plants is different to the growth in her body. AO1 [4 marks]

(a) *She could weight the plants or measure them with a ruler.*
(b) *These plants grow into a more branching shape.*

Answer grade: G–F. Both parts of the answer are correct but incomplete. (a) is a reasonable answer but for full marks it should state whether it is dry mass or wet mass that is being measured and the height of the plants. In (b), another difference could be the importance of cell elongation in plant growth, or the use of meristems.

Page 64 Selective breeding of chickens

Using selective breeding, farmers have produced chickens that lay large eggs.
(a) Describe the steps that would be used to do this. AO1 [3 marks]
(b) What might be an advantage of producing these chickens by genetic engineering rather than selective breeding? AO1 [1 mark]

(a) *Find chickens that lay the largest eggs and breed them.*
(b) *They might be healthier.*

Answer grade: G–F. In (a), the idea of choosing the chickens that lay the largest eggs is described but not the steps in selective breeding. In (b), the answer is too vague. For full marks, the idea of selecting the offspring and repeating this for generations needs to be described. Also, the idea of the speed of genetic engineering or the dangers of inbreeding with selective breeding needs to be stated.

Page 65 Taking cuttings

A garden centre wants to sell an attractively coloured geranium plant. They decide to produce many clones of the plant by taking cuttings.
(a) Suggest reasons why they chose this method. AO1 [1 mark]
(b) Give a possible disadvantage of mass producing plants by this method. AO1 [2 marks]

(a) *They can make lots of plants quickly.*
(b) *The plants might all die at once.*

Answer grade: G–F. (a) gives a correct reason but does not say what the plants will look like. For full marks the answer should state that the plants are all genetically identical to the attractive parent plant. (b) is also incomplete, as no reason is given for the death of the plants. It is important to comment on the lack of genetic variation and the resulting susceptibility to disease.

Page 67 Kite diagrams

Look at the kite diagram on page 67.

What do kite diagrams show?

Describe trends shown in the kite diagram. *AO1/2* [4 marks]

(a) The kite diagram shows the distribution of plants and animals found near a woodland path.

(b) It shows that the types of plants and animals change away from the path. Some cannot survive being walked on.

Answer grade: F–E. Only a partial answer is given in (a). It should refer to the numbers or percentage cover of organisms. (b) correctly states that some organisms could not survive being trampled on, but no individual organisms are identified.

Page 68 Plant growth in summer

(a) Explain why plants grow better in summer than in winter. *AO1* [2 marks]

(b) Suggest how this information is useful to gardeners. *AO2* [2 marks]

(a) Plants grow better in summer because it is usually warmer. There are no frosts to damage the plants.

(b) Gardeners should only grow their crops in summer.

Answer grade: F–E. For full marks in (a), explain that more light is available and that more light and warmth would speed up photosynthesis. An increased temperature would increase the rate of enzyme action. Link the faster rate of photosynthesis with more sugars (for energy) and proteins (for growth). (b) is too vague. State that the information would help to time seed planting, or to set up better conditions early in the year by using a greenhouse.

Page 69 Leaf adaptations

Explain why a plant needs chlorophyll and why it needs other pigments. *AO1* [4 marks]

Chlorophyll is needed to trap light energy in photosynthesis. Other pigments can use all forms of light.

Answer grade: F–E. This is too brief. For full marks, state that chlorophyll is a green pigment and that other pigments use different parts of the light spectrum, ensuring that a wide range of the light spectrum is used.

Page 70 Gases for plant growth

(a) Which gases does a green plant need?

(b) How is a leaf adapted to ensure a good supply of these gases? *AO1* [5 marks]

(a) A green plant needs carbon dioxide .

(b) A leaf is adapted by having a large surface area and stomata.

Answer grade: F–E. (a) does not identify that oxygen is needed. In (b), only two adaptations of leaves are given. For full marks, explain that stomata are small pores in leaf surfaces and describe the involvement of diffusion in the entry of gases.

Page 71 Transpiration

Explain how water from the soil reaches plant leaves. *AO1* [4 marks]

Water is absorbed by root hairs. It then travels up the stem to the leaves where it is needed.

Answer grade: F–E. The process is described but not explained. For full marks, state that water enters root hair cells by osmosis, as the water concentration is higher in the soil than in the root hair cells. Also state that the water is pulled up the stem inside the xylem cells and that this is caused by water evaporating from leaves (transpiration).

Page 72 Mineral deficiency

A fertiliser bag has a label containing NPK information. Explain why this information is important. *AO1* [4 marks]

NPK information describes the relative amounts of minerals present in the fertiliser. Minerals are needed for plant growth.

Answer grade: F–E. For full marks, state that: the letters NPK refer to the relative amounts of nitrates, phosphates and potassium compounds in the fertiliser, that each mineral is needed for a specific job (for example nitrate compounds are needed to make proteins) and that too small an amount will cause a mineral deficiency, for example a lack of potassium causes poor flower and fruit growth.

Page 73 Decay and decomposers

Describe an experiment to show that decay is caused by decomposers such as bacteria and fungi. *AO1* [4 marks]

Two samples of soil were collected. Sample A was heated but not burned. Sample B was not heated. Both samples were put in a flask containing lime water to see if the lime water turned milky.

Answer grade: F–E. For full marks, explain the results, such as the lime water in flask B turning milky, showing the presence of carbon dioxide from living organisms, and describe weighing the soil samples and the loss of weight in soil sample B, showing that decay had taken place.

Page 74 Hydroponics

(a) Describe how plants can be grown without soil. *AO1* [2 marks]

(b) Explain why this method is used in some areas of the world. *AO1* [3 marks]

(a) Plants such as tomato or lettuce plants can be grown without soil in containers in large greenhouses. The containers have a continuous supply of water and are easily harvested.

(b) It is used in hot and poor countries.

Answer grade: F–E. For full marks in (a) name the technique (hydroponics) and state that plants have a continuous supply of minerals and oxygen as well as water; that the amount of minerals can be regulated to avoid waste, and that the system is automated to save money. (b) is incorrect. Hydroponics can be used in areas of poor soil. It is an example of intensive farming.

C3 Improve your grade Chemical economics

Page 76 Change in rate of reaction

Limestone is reacted with excess hydrochloric acid in a flask. Explain why the mass of the flask and contents decreases and then the reaction stops. *AO1* [2 marks]

Carbon dioxide is given off so the mass decreases and the reaction stops because the reactants are used up.

Answer grade: G–E. This answer would gain a mark. For full marks, the student should have explained that the acid is in excess so the limestone is the limiting factor and is all used up at the end of the reaction.

Page 77 Temperature and rate of reaction

Explain the effect of changing the temperature on the rate of a chemical reaction. *AO1* [2 marks]

If the temperature is increased the reaction will go faster.

Answer grade: G–E. This is correct for one mark. For full marks the answer should have explained why this happens, i.e. because particles make more collisions or move faster at higher temperatures

Page 78 Surface area and rate of reaction

If the graph of two reactions, where the volume of gas is measured over time, shows one steep curve and one less steep curve, which graph shows the quicker reaction? *AO1* [1 mark]

The steeper curve shows the faster reaction.

Answer grade: G–E. This is correct. To improve your grade you may be asked to sketch another graph to show the effect of changing the surface area of reactants again on the rate of reaction.

Page 79 Conservation of mass

If 125 g of zinc carbonate made 81 g of zinc oxide and 44 g of CO_2, show that mass is conserved in this reaction.

$$ZnCO_3 \rightarrow ZnO + CO_2$$

AO2 [3 marks]

The products have a mass of 81 + 44 = 125 g which is the same as the mass of the reactant at the start.

Answer grade: G–E. This is correct. For full marks the answer should also have shown the conservation of mass using formula masses.

Page 80 Percentage yields and atom economy

A reaction has an atom economy of 86% and a percentage yield of 52%. What does this mean? *AO2* [2 marks]

This means that not all the atoms in the reactant will be converted to the product that is needed.

Answer grade: G–E. This is correct for atom economy. The percentage yield number means that only about half the expected yield is actually made and about half the product is lost. To improve your grade you may be asked to calculate the atom economy and the percentage yield of a reaction when you know which chemical is the desired product.

Page 81 Comparing energy in fuels

Three fuels heat the same mass of water when burnt under a calorimeter. Fuel A changes the temperature by 21 °C, Fuel B by 25 °C and Fuel C by 18 °C. Which fuel released the most energy? *AO2* [1 mark]

Fuel B released most energy.

Answer grade: G–E. This is correct. To improve your grade you may be asked to work out energy values from experiments using the formula:

energy transferred = $m \times c \times \Delta T$

Page 82 Drug development

What are the factors that affect the making of a drug? *AO1* [3 marks]

Labour costs, energy costs, raw materials.

Answer grade: G–E. This is correct. To improve your grade you may be asked to explain why it is often expensive to develop a new drug. Your answer should explain that research and testing can take up to ten years to find the one compound in thousands tried that will work.

Page 83 Structure of allotropes

Give two properties of diamond that are different from the properties of graphite. *AO1* [4 marks]

Diamond does not conduct electricity. It is not slippery and is transparent.

Answer grade: G–E. This is correct. To improve your grade you may be asked to explain why graphite can be used in pencils and why diamond can be used in jewellery.

Page 85 Arrangement of electrons

If a potassium atom has the electronic structure 2.8.8.1, how many shells do the electrons occupy? *AO2* [1 mark]

4

Answer grade: G–E. This is correct, as the pattern suggests. To improve your grade you may be asked to deduce the identity of the element from this type of electronic structure.

Page 86 Structure and bonding

Sodium chloride is made up of Na^+ particles and Cl^- particles. Are these atoms, ions or molecules? Is sodium chloride easier or harder to melt than magnesium oxide? *AO2* [2 marks]

These are ions. Sodium chloride is easier to melt than magnesium oxide.

Answer grade: G–E. This is correct. However, a better answer would have also said that magnesium oxide melts at a higher temperature than sodium chloride. To improve your grade you may be asked about the electrical conductivity of the compounds: you will need to know that magnesium oxide and sodium chloride both conduct electricity when they are molten.

Page 87 Elements in the periodic table

Chlorine is number 17 in the periodic table. Name two other elements that are in the same group as chlorine and two other elements that are in the same period. *AO2* [2 marks]

Bromine and iodine and phosphorus and sulfur.

Answer grade: G–E. This is correct but it would have been better to say which two elements were in the same group and which two were in the same period. To improve your grade you may be asked to explain that the group number is the same as the number of electrons in the outer shell, and to deduce to which group an element belongs from its electronic structure.

Page 88 Reactivity in alkali metals

Explain how group 1 elements react with water. *AO1* [4 marks]

Group 1 elements float on water and react vigorously. They give off hydrogen and make a hydroxide solution, which is alkaline.

Answer grade: G–E. This is correct and makes five points; enough for four marks. However, to improve your grade you may be asked to predict how rubidium will react with water and to construct the balanced symbol equation of any group 1 metal with water.

Page 89 Displacement reactions of halogens

If chlorine is bubbled into a solution of potassium iodide, then iodine and potassium chloride are made. Construct a word equation for this reaction. *AO1* [2 marks]

chlorine + potassium iodide → potassium chloride + iodine

Answer grade: G–E. This is correct and shows a displacement reaction. To improve your grade you may be asked to describe what you see during a reaction of this kind.

Page 90 Precipitation reactions

A reaction between a blue solution and sodium hydroxide produces a blue solid. What type of reaction is this? *AO1* [1 mark]

Precipitation

Answer grade: G–E. This is correct. To improve your grade may be asked to describe the use of sodium hydroxide to identify the presence of transition metal ions by recognising the coloured solids they make with it.

Page 91 Uses of metals

Explain which properties of metals are useful when making gardening tools. *AO2* [2 marks]

Strength, hardness and low density.

Answer grade: G–E. The three choices are relevant. To improve your grade you may be asked to explain why particular metals are suited to a given use from information given in data tables, e.g. a question may ask you to select a suitable metal for a garden tool given information about the malleability of different metals in a table.

Page 92 Testing water

Explain how you would test whether the ions in a sample of water were chloride or bromide ions. *AO1* [3 marks]

Add silver nitrate to a solution. If the precipitate is white, then chloride ions were in the solution. If it is a cream precipitate, then there were bromide ions in the solution.

Answer grade: G–E. The three points are identified. To improve your grade you may be asked to interpret data about the testing of water with silver nitrate and to construct word equations of the reactions of silver nitrate with halide ions.

Page 94 Calculating speed

A car travels 600 m in 20 s. Calculate its average speed.
AO1 [2 marks], AO2 [1 mark]

$$average\ speed = \frac{distance}{speed} = \frac{600}{20} = 30$$

Answer grade: F–E. The calculation is correct but there is no unit given for speed. For full marks, students should include m/s.

Page 95 Calculating acceleration

A cat is walking at a velocity of 0.5 m/s when it sees a mouse. The cat takes 2 s to accelerate to 5.5 m/s. Calculate its acceleration in m/s^2. *AO1 [1 mark], AO2 [2 marks]*

$$a = \frac{change\ in\ speed}{time} = \frac{5.5 - 0.5}{2}$$

Answer grade: F–E. The working is correct but the answer should be 2.5. For full marks, students should remember to double check their calculations.

Page 96 Stopping distance

Tim has a reaction time of 0.7 s. Calculate his thinking distance when he is travelling at 108 km/h. AO1 [1 mark], *AO2 [2 marks]*

$$108\ km/h = \frac{108 \times 1000}{3600} = 30\ m/s = 21\ m$$

Answer grade: F–E. The answer of 21 m is correct but the working is not fully shown. For full marks, students should remember to show all of their working out.

Page 97 Comparing power

Jan and Meera can both run up a flight of stairs in 8 s. Meera is more powerful than Jan. Explain why. *AO1 [1 mark]*

They both weigh differently.

Answer grade: G–F. The sentence is true but is not a full enough explanation. For full marks, students should explain that Meera's weight is greater than Jan's.

Page 98 Electric cars and pollution

Janice says that electric cars are better for the environment because they don't pollute it. James says that there is still pollution but not at the point of use. Explain what James means. *AO1 [2 marks]*

The electricity has to be made at a power station.

Answer grade: E–D. The sentence is true but is not a full enough explanation. For full marks, students should explain that power stations do pollute the environment.

Page 99 Safety features

Explain how crumple zones help to reduce injuries in a crash. *AO1 [3 marks]*

The crumple zone squashes up. It stops you more slowly.

Answer grade: F–E. The first sentence suggests the crumple zone is changing shape. The second has the idea of reducing momentum more slowly. For full marks, students should use the correct technical terms and mention the absorption of energy.

Page 100 Streamlining

Explain the effect a streamlined shape has on the motion of a car? *AO1 [2 marks]*

It goes faster.

Answer grade: G–F. The answer is a description of the effect, not an explanation. For full marks, students should explain that the drag is reduced which means that the car can reach a higher maximum speed. This will also improve the fuel consumption and reduce costs.

Page 101 Energy changes on a water chute

Janie is using the water chute at the swimming pool. The fall of the chute is 12 m. Janie's mass is 50 kg. The gravitational field strength is 10 N/kg. (a) What will be her kinetic energy at the bottom of the chute as she enters the pool? (b) Her speed as she enters the pool is twice her speed when she is 9 m above the pool. What is her kinetic energy when she is 9 m above the pool? *AO1 [2 marks], AO2 [2 marks]*

(a) KE = GPE = 50 × 10 × 12 = 6000 J
(b) 150 J

Answer grade: E–D. The answer to part (a) is correct, the working is shown and the correct unit given. The answer to part (b) is a factor of 10 too small. For full marks, students should explain that when speed increases by two, KE increases by four and show their working out.

Page 103 Static Charge

Connor is in the library walking on a nylon carpet. He touches a metal book shelf and receives an electric shock. Explain how he became charged and why he received a shock.
AO1 [3 marks]

He becomes charged and when he touches the shelf he discharges.

Answer grade: F–E. The basic ideas are correct but there is insufficient detail in the answer. Students should explain that walking on the nylon carpet causes friction on an insulator and that leads to charge building up on him. Then the charges flow to earth through the metal shelf causing a shock.

Page 104 Defibrillators

Explain how a human heart can be restarted using static electricity. *AO1* [3 marks]

Patient is given a shock to the heart to restore its beating. Other people have to stand clear.

Answer grade: G–F. A shock is used but this answer does not explain that the paddles have to be charged up from a high voltage supply first and then placed firmly on the chest to ensure good electrical contact. The charge is then discharged through the heart.

Page 105 Variable resistors

Explain how a variable resistor works and give an example of where one is used in the home. *AO1* [2 marks]

Variable resistors change the resistance. Longer wires add more resistance. They are used in lights.

Answer grade: F–E. The first part of the answer really just repeats the question. Students can improve this answer by explaining that the resistor will control the amount of current flowing. Longer wires will add more resistance and the current will decrease. These are used in dimmer switches on lighting circuits to control the amount of light.

Page 106 Uses of ultrasound

Ultrasound is used to treat kidney stones.
(a) Give two other medical uses of ultrasound
(b) Explain how kidney stones are treated. *AO1* [4 marks]

(a) scanning and sterilising instruments.
(b) high-powered ultrasound vibrations get rid of the kidney stones.

Answer grade: F–E.

(a) The answer needs to be more specific by saying which part of the body is scanned, e.g. prenatal scanning. Also ultrasound is used for cleaning but not sterilisation. Cleaning teeth or measuring blood flow would be better answers.

(b) The ultrasound will break down the kidney stones but then the pieces must be excreted normally.

Page 107 Nuclear radiation

What is an alpha particle? *AO1* [1 mark]

Alpha is a type of radioactivity from the nucleus which is weaker than beta or gamma.

Answer grade: G. The question is for the constituents of an alpha particle not for any properties. An alpha is two neutrons and two protons (or a helium nucleus).

Page 108 Tracers

How are radioactive sources used to find blockages in underground pipes? *AO1* [3 marks]

The radioactivity will not be detected after the leak.

Answer grade: F–E. The answer lacks detail. To improve the answer students should mention that radioactive material is put into the pipe and its progress can be tracked with a detector above ground. The blockage is then shown by a reduction in radioactivity after the point of blockage.

Page 109 Medical tracers

Why is iodine-123 used as a tracer in medicine? *AO1* [3 marks]

A tracer is a substance put into the body to find out if there is a problem. Iodine is absorbed by the thyroid gland and can show up any problems there.

Answer grade: F–E. The answer gives two correct statements but not full explanations. For full marks, students should explain that a tracer is a radioisotope that will emit radiation that can be detected from the outside of the body. Also explain that since it will be emitting potentially harmful radiation it must have a short half-life so that the patient's exposure is limited.

Page 110 Nuclear fission

Explain how nuclear fission produces electricity.
AO1 [3 marks]

The uranium in a nuclear reactor splits and starts a chain reaction which gives off energy. The energy is used to generate electricity.

Answer grade: F–E. These statements are correct but do not give a clear enough explanation of the process. For full marks, students should explain that the splitting of the uranium nuclei sets off a chain reaction which creates a lot of heat. This heat is used to boil water to produce steam at high pressure which is used to turn turbines. The turbines in turn spin the generators thus generating electricity.

Understanding the scientific process

As part of your Science assessment, you will need to show that you have an understanding of the scientific process – How Science Works.

This involves examining how scientific data is collected and analysed. You will need to evaluate the data by providing evidence to test ideas and develop theories. Some explanations are developed using scientific theories, models and ideas. You should be aware that there are some questions that science cannot answer and some that science cannot address.

Collecting and evaluating data

You should be able to devise a plan that will answer a scientific question or solve a scientific problem. In doing so, you will need to collect data from both primary and secondary sources. Primary data will come from your own findings – often from an experimental procedure or investigation. While working with primary data, you will need to show that you can work safely and accurately, not only on your own but also with others.

Secondary data is found by research, often using ICT – but do not forget books, journals, magazines and newspapers are also sources. The data you collect will need to be evaluated for its validity and reliability as evidence.

Presenting information

You should be able to present your information in an appropriate, scientific manner. This may involve the use of mathematical language as well as using the correct scientific terminology and conventions. You should be able to develop an argument and come to a conclusion based on recall and analysis of scientific information. It is important to use both quantitative and qualitative arguments.

Changing ideas and explanations

Many of today's scientific and technological developments have both benefits and risks. The decisions that scientists make will almost certainly raise ethical, environmental, social or economic questions. Scientific ideas and explanations change as time passes and the standards and values of society change. It is the job of scientists to validate these changing ideas.

A

ABS braking system, known as advance braking system, which helps to control a skidding car   99

acceleration a measurement of how quickly the speed of a moving object changes (when speed is in m/s the acceleration is in m/s^2)   95, 96, 99, 100

acid solution with a pH of less than 7   28, 36, 37

acid rain rain water which is made more acidic by pollutant gases   19, 23, 24, 34

acrosome part of the sperm that contains enzymes   61

active immunity you have immunity if your immune system recognises a pathogen and fights it   6

active site the place on an enzyme where the substrate molecule binds   59

adaptations features that organisms have to help them survive in their environment   17, 18

aerobic respiration respiration that involves oxygen   60

air bags cushions which inflate with gas to protect people in a vehicle accident   99

air resistance the force exerted by air to any object passing through it   100, 101

alcohol substance made by the fermentation of yeast   4, 8, 28

alkali metals very reactive metals in group 1 of the periodic table, e.g. sodium   88, 89

alkalis substances which produce OH– ions in water   36, 37, 88

alkanes a family of hydrocarbons found in crude oil with single covalent bonds, e.g. methane   22, 25

alkenes a family of hydrocarbons with one double covalent bond between carbon atoms   22, 25

allele inherited characteristics are carried as pairs of alleles on pairs of chromosomes. Different forms of a gene are different alleles   11

allotropes different forms of the same element   83

alloy mixture of two or more metals – used to make coins, for example   33, 34

alpha particles radioactive particles which are helium nuclei – helium atoms without the electrons (they have a positive charge)   107, 108, 109

alternating current or voltage an electric current that is not a one-way flow   50

amalgam an alloy which contains mercury   33

amino acids small molecules from which proteins are built   5, 59

amplitude the amplitude of a wave is the maximum displacement of a wave from its rest position   42, 45, 106

anaerobic respiration respiration without using oxygen   60

analogue signal a signal that shows a complete range of frequencies; sound is analogue   45

anode electrode with a positive charge   33, 38

antibiotic therapeutic drug acting to kill bacteria which is taken into the body   6, 18

antibody protein normally present in the body or produced in response to an antigen which it neutralises, thus producing an immune response   6

antigen any substance that stimulates the production of antibodies – antigens on the surface of red blood cells determine blood group   6

antiviral drug therapeutic drug acting to kill viruses   6

arteries blood vessels that carry blood away from the heart   4, 62

asexual reproduction reproduction involving only one parent   61, 65

asteroid composed of rock or metallic material orbiting the Sun in a region between Mars and Jupiter   55

atmosphere mixture of gases above the lithosphere, mainly nitrogen and oxygen   24, 46, 47, 49, 51, 54, 104

atom the basic 'building block' of an element which cannot be chemically broken down   25, 53, 79, 80, 83, 85, 86, 87, 88, 89, 91, 103, 107, 109

atom economy a way of measuring the amount of atoms that are wasted or lost when a chemical is made   80

atomic number the number of protons found in the nucleus of an atom   85

attract move towards, for example, unlike charges attract   86, 87, 103, 104

auxin a type of plant hormone   10

average speed total distance travelled divided by the total time taken for a journey   94

axon part of neurone that carries nerve impulse   7

B

background radiation ionising radiation from space and rocks, especially granite, that is around us all the time but is at a very low level   108

bacteria single-celled micro-organisms which can either be free-living organisms or parasites (they sometimes invade the body and cause disease)   4, 6, 14, 15, 18, 53, 61, 63, 64, 73, 92

balanced symbol equation a symbolic representation showing the kind and amount of the starting materials and products of a reaction   27, 35, 68, 88, 89

barium chloride testing chemical for sulfates in water   92

base a substance that will react with acids   36, 58

batch process a process used to make small fixed amounts of substances, like medicines, with a clear start and finish   82

battery two or more electrical cells joined together   34, 98

beta particles particles given off by some radioactive materials (they have a negative charge)   107, 109

Big Bang the event believed by many scientists to have been the start of the universe   56

binocular vision the ability to maintain visual focus on an object with both eyes, creating a single visual image   7, 17

binomial system the scientific way of naming an organism   13

biodegradable a biodegradable material can be broken down by micro-organisms   26

biodiversity range of different living organisms in a habitat   67

biofuels fuels made from plants – these can be burned in power stations   98

biological catalyst molecules in the body that speed up chemical reactions   59

biological control a natural predator is released to reduce the number of pests infesting a crop   74

biomass waste wood and other natural materials which are burned in power stations   14

bitumen thick tar-like substance that does not boil in a fractionating column   22

black hole a region of space from which nothing, not even light, can escape   54, 56

blood pressure force with which blood presses against the walls of the blood vessels   4

blood sugar level amount of glucose in the blood   9, 59

body mass index (BMI) measure of someone's weight in relation to their height   5

boiling point temperature at which the bulk of a liquid changes into a gas   22, 40, 82, 83, 88, 91

boron control rods rods that are raised or lowered in a nuclear reactor to control the rate of fission 110

braking distance distance travelled while a car is braking   96

brass an alloy which contains copper and zinc   33

bromine an orange, corrosive halogen, used to test alkenes   25, 89

buckminsterfullerene a very stable sphere of 60 carbon atoms joined by covalent bonds. An allotrope of carbon   83

C

cancer life-threatening condition where body cells divide uncontrollably   6, 47, 53, 109

capillaries small blood vessels that join arteries to veins   62

captive breeding breeding a species in zoos to maintain the wild population   20

carbon a very important element, carbon is present in all living things and forms a huge range of compounds with other elements   15, 22, 83, 85, 87

carbon cycle a natural cycle through which carbon moves by respiration, photosynthesis and combustion in the form of carbon dioxide   24

carbon dioxide (CO_2) gas present in the atmosphere at a low percentage but important in respiration, photosynthesis and combustion; a greenhouse gas which is emitted into the atmosphere as a by-product of combustion   9, 15, 19, 23, 24, 27, 32, 51, 53, 60, 62, 68, 69, 70, 78, 85, 86, 87, 90, 97, 98

carbon monoxide poisonous gas made when fuels burn in a shortage of oxygen   4, 8, 23, 24

carbon-14 radioactive isotope of carbon   85, 108

carotene plant pigment involved in photosynthesis   64

catalyst a chemical that speeds up a reaction but is not changed or used up by the reaction   22, 25, 35, 78, 83, 90

catalytic converters boxes fitted to vehicle exhausts which reduce the level of nitrogen oxides and unburnt hydrocarbons in the exhaust fumes   24

cathode electrode in a battery with a negative charge   33, 38

cell differentiation when cells become specialised   61, 63

cell membrane layer around a cell which helps to control substances entering and leaving the cell   11, 70

cement the substance made when limestone and clay are heated together   32, 51

central nervous system (CNS) collectively the brain and spinal cord   7

CFCs gases which used to be used in refrigerators and which harm the ozone layer   19, 47

chain reaction a reaction where the products cause the reaction to go further or faster, e.g. in nuclear fission   110

charge(s) a property of matter charge exists in two forms, positive and negative, which attract each other   85, 86, 103, 104

chemical properties the characteristic reactions of substances   87, 91

chlorination addition of chlorine to water supplies to kill micro-organisms   92

chlorophyll pigment found in plants which is used in photosynthesis (gives green plants their colour)   68, 69

cholesterol fatty substance which can block blood vessels   4

chromatography a method for splitting up a substance to identify compounds and check for purity   82

chromosomes thread-like structures in the cell nucleus that carry genetic information   11, 58, 61

circuit-breakers resettable fuses   105

clone genetically identical copy   65

cold fusion attempts to produce fusion at normal room temperature that have not been validated since other scientists could not reproduce the results   110

collagen protein used for support in animal cells   59

colloid a liquid with small particles dispersed throughout it, forming neither solution nor sediment   29

combustion process where fuels react with oxygen to produce heat   15, 23, 24, 81

comets lumps of rock and ice found in space – some orbit the Sun   54, 55

community all the plants and animals living in an ecosystem, e.g. a garden   67

complete combustion when fuels burn in excess of oxygen to produce carbon dioxide and water only   23

composite material a material which consists of identifiably different substances   32

compost dead and decaying plant material   73, 74

compound two or more elements which are chemically joined together, e.g. H_2O 15, 24, 25, 33, 79, 82, 85, 88, 90, 92

compressions particles pushed together, increasing pressure   106

computer modelling using a computer to 'model' situations to see how they are likely to work out if you do different things   99

concentration the amount of chemical dissolved in a certain volume of solution   36, 73, 77

concrete a form of artificial stone   32, 53

conservation a way of protecting a species or environment   20

conservation of mass the total mass of reactants equals the total mass of products formed   79

core the centre part of the a planet, made of iron   31, 55, 56

corrode to lose strength due to chemical attack   34

cosmic rays radiation from space that contributes to background radiation   108

count rate average number of nuclei that decay every second   107

covalent bond bond between atoms where an electron pair is shared   25, 83, 87

cracking the process of making small hydrocarbon molecules from larger hydrocarbon molecules using a catalyst   22, 35

critical angle angle at which a light ray incident on the inner surface of a transparent glass block just escapes from the glass   43

crop rotation system of growing crops in sequence   74

cross-sectional area the area displaced by a moving object   100

crude oil black material mined from the Earth from which petrol and many other products are made   22, 26, 34

crumple zones areas of a car that absorb the energy of a crash to protect the centre part of the vehicle   99

crust surface layer of the Earth made of tectonic plates   31, 32, 55

current flow of electrons in an electric circuit   49, 50, 52, 53, 105, 108

D

decay to rot   15, 51, 73, 107

deceleration a measurement of how quickly the speed of a moving object decreases   95

decolourise turn from a coloured solution to a colourless solution   25

decomposer organisms that break down dead animals and plants   14, 15, 73

decomposed chemically broken down   26, 27

defibrillator machine which gives the heart an electric shock to start it beating regularly   104

deforestation removal of large area of trees   51

dehydration result of body losing too much water   5, 9

density the density of a substance is found by dividing its mass by its volume   33, 34, 42, 55, 56, 91, 109

depressant a drug that slows down the working of the brain   8

detritivore an organism that eats dead material, e.g. an earthworm   73

detritus the dead and semi-decayed remains of living things   73

diastolic pressure the lowest point that your blood pressure reaches as the heart relaxes between beats   4

diet what a person eats   4, 5, 6, 9

diffraction a change in the directions and intensities of a group of waves after passing by an obstacle or through an opening whose size is approximately the same as the wavelength of the waves   42

diffuse when particles diffuse they spread out   69, 70, 71

digital signal a signal that has only two values – on and off   43, 45, 46

diploid cells that have two copies of each chromosome   61

displacement reaction chemical reaction where one element displaces or 'pushes out' another element from a compound   89

displayed formula when the formula of a chemical is written showing all the atoms and all the bonds   25

disposal getting rid of unwanted substances such as plastics   26

distance–time graph a plot of the distance moved against the time taken for a journey   94

distillation the process of evaporation followed by condensation   22, 28

DNA molecule found in all body cells in the nucleus – its sequence determines how our bodies are made (e.g. do we have straight or curly hair), and gives each one of us a unique genetic code   53, 58, 61, 65

dominant allele/characteristic an allele that will produce the characteristic if present   11

double covalent bond covalent bond where each atom shares two electrons with the other atom   22, 25

drag energy losses caused by the continual pushing of an object against the air or a liquid   100

dummies used in crash testing to learn what would happen to the occupants of a car in a crash   99

dynamo a device that converts energy in movement into energy in electricity   50

E

EAR for protein estimated average daily requirement of protein in diet   5

earth wire the third wire in a mains cable which connects the case of an appliance to the ground so that the case cannot become charged and cause an electric shock   105

earthed (electrically) connected to the ground (at 0V)   103, 104

efficiency ratio of useful energy output to total energy input; can be expressed as a percentage   4, 41, 50, 74, 98

egestion the way an animal gets rid of undigested food waste called faeces   14

electrical conductors materials which let electricity pass through them   34, 91, 103

electrolysis when an electric current is passed through a solution which conducts electricity   33, 38

electrolyte the liquid in which electrolysis takes place   33

electromagnet a magnet which is magnetic only when a current is switched on   91

electromagnetic spectrum electromagnetic waves ordered according to wavelength and frequency – ranging from radio waves to gamma rays   42, 44, 46

electron shells the orbit around the nucleus likely to contain the electron   85

electronic structure the number of electrons in sequence that occupy the shells, e.g. the 11 electrons of sodium are in sequence 2.8.1   85, 89

electrons small particles within an atom that orbit the nucleus (they have a negative charge)   53, 56, 85, 86, 87, 88, 89, 91, 103, 107, 109

electrostatic attraction attraction between opposite charges, e.g. between Na^+ and Cl^- 86

elements substances made out of only one type of atom   34, 79, 83, 85, 87, 89, 90

elliptical orbit a path that follows an ellipse – which looks a bit like a flattened circle   54

endangered a species where the numbers are so low they could soon become extinct   20

endothermic reaction chemical reaction in which heat is taken in   81

energy the ability to 'do work', for example the human body needs energy to function   5, 9, 13, 14, 23, 27, 29, 40, 41, 49, 50, 51, 52, 58, 60, 68, 69, 81, 92, 97, 98, 99, 101, 105, 106, 109, 110

enriched uranium uranium containing more of the U-235 isotope than occurs naturally   110

enzymes biological catalysts that increase the speed of a chemical reaction   9, 59, 61, 68

essential elements the three elements, nitrogen, phosphorus and potassium that are essential for the growth of plants   37

eutrophication when waterways become too rich with nutrients (from fertilisers) which allows algae to grow wildly and use up all the oxygen   37

evaporation when a liquid changes to a gas, it evaporates   9, 28, 29, 37, 38, 51, 71, 80

evolution the gradual change in organisms over millions of years caused by random mutations and natural selection   13, 18

excretion the process of getting rid of waste from the body   14

exhaust gases gases discharged into the atmosphere as a result of combustion of fuels   97

exothermic reaction chemical reaction in which heat is given out   81

explosion a very fast reaction making large volumes of gas   31, 56, 78, 103

exponential growth the ever-increasing growth of the human population   19

extinct when all members of a species have died out   18, 20

F

fertilisation when a sperm fuses (joins with) an egg   11, 61

fertiliser chemical put on soil to increase soil fertility and allow better growth of crop plants   19, 35, 37, 72, 73, 74, 80, 82, 90, 92, 104

filtration the process of filtering river or ground water to purify it for drinking water   80, 92

finite resource resources such as oil that will eventually run out   19, 22, 34

fission splitting apart, especially of large radioactive nuclei such as uranium   110

flame test test where a chemical burns in a Bunsen flame with a characteristic colour – tests for metal ions   88

force a push or pull which is able to change the velocity or shape of a body   87, 96, 97, 99, 100

fossil fuels fuels such as coal, oil and gas   15, 19, 22, 24, 50, 51, 52, 98

free-fall a body falling through the atmosphere without an open parachute   100

frequency the number of waves passing a set point per second   42, 46, 50, 106

friction energy losses caused by two or more objects rubbing against each other   100, 101, 103

fuel consumption the distance travelled by a given amount of fuel, e.g. in km/100 litres   97, 98

fuel rods rods of enriched uranium produced to provide fuel for nuclear power stations   110

fullerenes cage-like carbon molecules containing many carbon atoms, e.g. buckyballs   83

fungicide chemical used to kill fungi   74

fuse(s) a special component in an electric circuit containing a thin wire which is designed to melt if too much current flows through it, breaking the circuit   105

fusion the joining together of small nuclei, such as hydrogen isotopes, at very high temperatures with the release of energy   110

G

gametes the male and female sex cells (sperm and eggs)   11, 61

gamma rays ionising electromagnetic waves that are radioactive and dangerous to human health – but useful in killing cancer cells   42, 107, 108, 109

gene section of DNA that codes for a particular characteristic   11, 18, 58, 59, 61, 64

gene therapy medical procedure where a virus is used to 'carry' a gene into the nucleus of a cell (this is a new treatment for genetic diseases)   64

generator device that converts rotational kinetic energy to electrical energy   50, 110

genetic engineering transfer of genes from one organism to another   64, 65

geotropism a plant's growth response to gravity   10

giant ionic lattice sodium chloride forms a lattice, also called a giant ionic structure   86

global warming the increase in the Earth's temperature due to increases in carbon dioxide levels   19, 49, 51, 53, 92

gradient rate of change of two quantities on a graph; change in y/change in x   76, 94, 95

granite an igneous rock containing low levels of uranium   32, 108

graphite a type of carbon used as a moderator in a nuclear power station   83, 110

gravitational field strength the force of attraction between two masses   97, 101

gravitational potential energy the energy a body has because of its position in a gravitational field, e.g. an object   101

gravity an attractive force between objects (dependent on their mass)   10, 97, 100

greenhouse gas any of the gases whose absorption of infrared radiation from the Earth's surface is responsible for the greenhouse effect, e.g. carbon dioxide, methane, water vapour 51, 97

group 1 metals metals in the group 1 of the periodic table, e.g. lithium, sodium and potassium 88

group 7 elements non-metals in group 7 of the periodic table, e.g. fluorine, bromine and iodine 89

H

Haber process industrial process for making ammonia 35, 90

habitat where an organism lives, e.g. the worm's habitat is the soil 13, 14, 16, 20, 38, 67, 74

haemoglobin chemical found in red blood cells which carries oxygen 5, 59, 62

half-life average time taken for half the nuclei in a radioactive sample to decay 107

hallucinogen a drug, like LSD, that gives the user hallucinations 8

halogens reactive non-metals in group 7 of the periodic table , e.g. chlorine 89

haploid cells that have only one copy of each chromosome 61

heat stroke result of body being too hot; skin is cold, pulse is weak 9

helium second element in periodic table; an alpha particle is a helium nucleus 56, 85, 107

herbicide chemical used to kill weeds 64, 74

hertz (Hz) units for measuring wave frequency 42, 50, 106

hormones chemicals that act on target organs in the body (hormones are made by the body in special glands) 9, 59, 62

hydrated iron(III) oxide the chemical name for rust 34

hydroponics growing plants in mineral solutions without the need for soil 74

hypothermia a condition caused by the body getting too cold, which can lead to death if untreated 9

I

igneous rock rock which has formed when liquid rock has solidified 31

inbreeding breeding closely related animals 64

indicator species organisms used to measure the level of pollution in water or the air 19

infrared waves non-ionising waves that produce heat – used in toasters and electric fires 41, 43, 44, 45, 51

insecticide a chemical that can kill an insect 74, 104

instantaneous speed the speed of a moving object at one particular moment 94

insulation a substance that reduces the movement of energy; heat insulation in the loft of a house slows down the movement of warmth to the cooler outside 17, 41, 103

insulin hormone made by the pancreas which controls the level of glucose in the blood 9, 59, 64

intensive farming farming that uses a lot of artificial fertilisers and energy to produce a high yield per farm worker 74

interference waves interfere with each other when two waves of different frequencies occupy the same space; interference occurs in light and sound and can produce changes in intensity of the waves 46

intermolecular force force between molecules 87

iodine radioactive isotopes of iodine are used in diagnosing and treating thyroid cancer 89, 109

ionic bond a chemical bond between two ions of opposite charges 86, 87

ionisation the formation of ions (charged particles) 53, 107, 108, 109

ionises adds or removes electrons from an atom, leaving it charged 53, 107, 108

ions charged particles (can be positive or negative) 36, 86, 90, 92

isotopes atoms with the same number of protons but different numbers of neutrons 85, 108, 110

J

joule unit of work done and energy 40, 97, 101

K

kilowatt 1000 watts 52, 97

kilowatt-hour unit of electrical energy equal to 3 600 000 J 52

kinetic energy the energy that moving objects have 44, 77, 98, 99, 101

kite diagram method of displaying results from a transect line 67

kwashiorkor an illness caused by protein deficiency due to lack of food. Sufferers often have swollen bellies caused by retention of fluid in the abdomen 5

L

laser a special kind of light beam that can carry a lot of energy and can be focused very accurately; lasers are often used to judge the speed of moving objects or the distance to them 43, 45, 104

lead heaviest element having a stable isotope; all isotopes of the elements above it in the periodic table are unstable 33, 34, 53, 92, 108

light-year a unit of distance equal to the distance light travels through space in one year 54

limestone a sedimentary rock, made of calcium carbonate 15, 32

limiting reactant chemical used up in a reaction that limits the amount of product formed 76

lithosphere the cold rigid outer part of the Earth which includes the crust and the upper part of the mantle 31

live wire carries a high voltage into and around the house 105

longitudinal (wave) wave in which the vibrations are in the same direction as the direction in which the wave travels 47, 106

M

magma molten rock found below the Earth's surface 31

malleable bendable 34, 91

mammal animals that have fur and produce milk for their young 11, 60, 61

mantle semi-liquid layer of the Earth beneath the crust 31

marble a metamorphic rock, made of calcium carbonate 32

mass describes the amount of something; it is measured in kilograms (kg) 40, 54, 76, 77, 78, 79, 80, 81, 98, 100, 101, 107

mass number number of protons and neutrons in a nucleus 85

meiosis cell division that results in haploid cells 61

melanin the group of naturally occurring dark pigments, especially the pigment found in skin, hair, fur, and feathers 47

melting point temperature at which a solid changes into a liquid 33, 82, 83, 86, 88, 91

meristem tips of roots and shoots where cell division and elongation takes place 63

metal halide a compound of a halogen and a metal, e.g. potassium bromide 89

metallic bonding the bonding between close-packed metal ions due to delocalised electrons 91

metallic properties the physical properties specific to a metal, such as lustre and electrical conductivity 90

metals solid substances that are usually lustrous, conduct electricity and form ions by losing electrons 86, 87, 88, 90, 91, 104

meteors bright flashes in the sky caused by rocks burning in Earth's atmosphere 54

microbes tiny microscopic organisms 53, 92

microorganism very small organism (living thing) which can only be viewed through a microscope – also known as a microbe 73

microwaves non-ionising waves used in satellite and mobile phone networks – also in microwave ovens 42, 44

minerals natural solid materials with a fixed chemical composition and structure, rocks are made of collections of minerals; mineral nutrients in our diet are things like calcium and iron, they are simple chemicals needed for health   68, 71, 72, 74

mitochondria structures in a cell where respiration takes place   58, 61, 63

mitosis cell division that results in genetically identical diploid cells   61

moderator material used to slow down neutrons in a nuclear power station   110

molecule two or more atoms which have been chemically combined   58, 59, 60, 70, 83, 86, 87, 89

molten liquid a solid that has just melted, usually referring to rock, ores, metals or salts with very high melting points   86

momentum the product of mass and velocity   99

Morse code a code consisting of dots and dashes that code for each letter of the alphabet   43

motor neurone nerve cell carrying information from the central nervous system to muscles   7

multicellular organism organisms made up of many specialised cells   61

mutation where the DNA within cells have been altered (this happens in cancer)   11, 59

mutualism a relationship in which both organisms benefit   16

N

nanometre units used to measure very small things (one billionth of a metre)   83

nanotube carbon atoms formed into a very tiny tube   83

National Grid network that carries electricity from power stations across the country (it uses cables, transformers and pylons)   50, 52

natural selection process by which 'good' characteristics that can be passed on in genes become more common in a population over many generations ('good' characteristics mean that the organism has an advantage which makes it more likely to survive)   18

near-Earth object asteroid, comet or large meteoroid whose orbit crosses Earth's orbit   55

negative ion an ion made by an atom gaining electrons   86

neutral a neutral substance has a pH of 7   37, 85

neutral wire provides a return path for the current in a mains supply to a local electricity substation   105

neutralisation reaction between H^+ ions and OH^- ions (acid and base react to make a salt and water)   36, 37

neutrons small particle which does not have a charge found in the nucleus of an atom   56, 85, 107, 109, 110

newtons unit of force (abbreviated to N)   96

non-metals substances that are dull solids, liquids or gases that do not conduct electricity and form ions by gaining electrons   34, 86, 87

non-renewable energy energy which is used up at a faster rate than it can be replaced e.g. fossil fuels   22, 50

nuclear power stations power stations using the energy produced by nuclear fission to generate heat   52, 53, 110

nuclear transfer type of cloning that involves taking a nucleus from a body cell and placing it into an egg cell   65

nucleus central part of an atom that contains protons and neutrons   53, 58, 61, 62, 63, 65, 85, 86, 103, 107, 109

O

obesity a medical condition where the amount of body fat is so great that it harms health   5

ohms units used to measure resistance to the flow of electricity   105

optical fibre a flexible optically transparent fibre, usually made of glass or plastic, through which light passes by successive internal reflections   43, 45

optimum conditions the conditions under which a reaction works most effectively   35

optimum temperature the temperature range that produces the best reaction rate   9

oscilloscope a device that displays a line on a screen showing regular changes (oscillations) in something. an oscilloscope is often used to look at sound waves collected by a microphone   50

osmosis a type of diffusion which depends on the presence of a partially-permeable membrane that allows the passage of water molecules but large molecules such as glucose   70, 71

oxidation a chemical reaction in which a substance gains oxygen and/or loses electrons   34

ozone layer layer of the Earth's atmosphere that protects us from ultraviolet rays   19, 47

P

paddle shift controls controls attached to the steering wheel of a car so that the driver can use them without taking their eyes off the road   99

paddles charged plates in a defibrillator that are placed on the patient's chest   104

painkiller a drug that stops nerve impulses so pain is not felt   8

palisade cells tightly packed together cells found on the upper side of a leaf   69

parasite organism which lives on (or inside) the body of another organism   6, 14, 16

partially-permeable membrane a membrane that allows some small molecules to pass through but not larger molecules   70

pathogen harmful organism which invades the body and causes disease   6

payback time the time it takes for the original cost outlay to be recovered in savings   41

percentage yield the percentage of actual product made in a chemical reaction compared to the amount which ideally could be made   80

performance enhancer a drug used to improve performance in a sporting event   8

period a row in the periodic table   85, 87

periodic table a table of all the chemical elements based on their atomic number   79, 85, 87, 90

peripheral nervous system (PNS) network of nerves leading to and from the brain and spinal cord   7

pesticide residue unwanted residues sometimes found in water contaminated by local pesticide use   92

petrol volatile mixture of hydrocarbons used as a fuel   22, 23, 98

pH scale scale in which acids have a pH below 7, alkalis a pH of above 7 and a neutral substance a pH of 7   15, 36, 59

pharmaceuticals medical drugs   82

phase fraction of a complete wave that one wave disturbance is different to another   43

phloem specialised transporting cells which form tubules in plants to carry sugars from leaves to other parts of the plant   71

photocell a device which converts light into electricity   49

photosynthesis process carried out by green plants where sunlight, carbon dioxide and water are used to produce glucose and oxygen   10, 14, 15, 24, 59, 68, 69, 71, 72, 73

phototropism a plant's growth response to light   10

physical property property that can be measured without changing the chemical composition of a substance, e.g. hardness   83, 91

pitch whether a sound is high or low on a musical scale   106

plant hormones hormones that control various plant processes such as growth and germination   10

plaque build up of cholesterol in a blood vessel (which may block it)   4

plasma yellow liquid found in blood   62

plasmolysis the shrinking of a plant cell due to loss of water, the cell membrane pulls away   70

plutonium a radioactive metal often formed as a bi-product from a nuclear power station – sometimes used as a nuclear fuel   52, 53

Glossary

pollination the process of transferring pollen from one plant to another   16

pollutants unwanted residues found that can sometimes cause damage   19, 24, 92

pollute contaminate or destroy the environment   49, 97, 98, 104

pollution contaminating or destroying the environment as a result of human activities   19, 23, 24, 33, 34, 49, 98

polymer a number of short-chained molecules joined together to form a long-chained molecule   25

population group of organisms of the same species in a habitat   19, 37, 67, 74

positive ion an ion made by an atom losing electrons   86, 88

potential difference another word for voltage (a measure of the energy carried by the electric current)   105

power the rate that a system transfers energy, power is usually measured in watts (W); electric power = voltage × current   52, 97, 105

power station facility that generates electricity on a large scale   50, 52, 98, 110

power transmission transmission of electricity   91

precipitate solid formed in a solution during a chemical reaction   90, 92

precipitation reaction chemical test in which a solid precipitate is formed – tests for metal ions   90, 92

predator animal which preys on (and eats) another animal   16, 17, 74

pressure wave vibrating particles in a longitudinal wave creating pressure variations   47, 106

prey animals which are eaten by a predator   16

probe unmanned space vehicle designed to travel beyond Earth's orbit   54

producers organisms in a food chain that make food using sunlight   14

product molecules produced at the end of a chemical reaction   76, 77, 79, 80, 82, 89

protons small positive particles found in the nucleus of an atom   53, 56, 85, 86, 107

P wave longitudinal seismic wave capable of travelling through solid and liquid parts of the Earth   47

R

radiation thermal energy transfer which occurs when something hotter than its surroundings radiates heat from its surface   41, 42, 44, 49, 51, 53, 54, 59, 107, 108, 109

radio waves non-ionising waves used to broadcast radio and TV programmes   42, 46

radioactive material which gives out radiation   107, 108, 109

radioactive waste waste produced by radioactive materials used at nuclear power stations, research centres and some hospitals   53, 108, 110

radiotherapy using ionising radiation to kill cancer cells in the body   53, 109

random having no regular pattern   11, 70, 107

rarefactions particles further apart than usual, decreasing pressure   106

rate of reaction the speed with which a chemical reaction takes place   76, 77, 78

reactants chemicals which are reacting together in a chemical reaction   76, 77, 78, 79, 80, 82, 89

reaction time the time it takes for a driver to step on the brake after seeing an obstacle   8, 96

receiver device which receives waves, e.g. a mobile phone   42, 44

recessive allele/characteristic two recessive alleles needed to produce the characteristic   11

recharging battery being charged with a flow of electric current   98

recycle to reuse materials   15, 26, 33, 34, 35

red blood cells blood cells which are adapted to carry oxygen   6, 62

reduction a chemical reaction in which a substance loses oxygen and/or gains electrons   33

reflected radiation rebounding off a surface   42, 43, 44, 45, 46

reflex a muscular action that we take without thinking about   7

refraction when a light ray travelling through air enters a glass block and changes direction   7, 42, 43, 45, 46

reinforced concrete concrete with steel rods or mesh running through it   32

relative atomic mass the mass of an atom compared to $\frac{1}{12}$ of a carbon atom   79

relative velocity vector difference between the velocities of two objects   95

renewable energy energy that can be replenished at the same rate that it's used up e.g. biofuels   49

repel move away, for example, like charges repel   91, 103, 104

reptile cold blooded vertebrate having an external covering of scales or horny plates   13

reservoir a water resource where large volumes of water are held   92

resistance measurement of how hard it is for an electric current to flow through a material   91, 105

respiration process occurring in living things where oxygen is used to release the energy in foods   9, 14, 15, 24, 51, 58, 59, 60, 68, 72

respiratory quotient (RQ) the result of dividing carbon dioxide produced by oxygen used, can be used to determine the types of molecule used in respiration   60

rheostat a variable resistor   105

rust the substance made when iron corrodes, hydrated iron(III) oxide   34

S

salt the substance formed when any acid reacts with a base   34, 36, 38

satellite a body orbiting around a larger body; communications satellites orbit the Earth to relay television and telephone signals   44

sea water water containing high levels of dissolved salts making it undrinkable   38

seat belts harness worn by occupants of motor vehicles to prevent them from being thrown about in a collision   99

sedimentation a process during water purification where small solid particles are allowed to settle   92

seismic wave vibration transmitted through the Earth   31

selective breeding process of breeding organisms with the desired characteristics   64

sensor device that detects a change in the environment   45

sensory neurone nerve cell carrying information from receptors to central nervous system   7

sex chromosomes a pair of chromosomes that determine gender, XX in female, XY in male   11

shock occurs when a person comes into contact with an electrical energy source so that electrical energy flows through a portion of the body   26, 103, 104, 105

silver nitrate a chemical used for testing halide ions in water   92

single covalent bond bond between atoms where each the atoms share an electron pair   25

smoke detector device to detect smoke, some forms of which contain a source of alpha radiation   108

Solar System the collection of planets and other objects orbiting around the Sun   54, 55

solar power energy provided by the Sun   98

solder an alloy which contains lead and tin   33

soluble a soluble substance can dissolve in a liquid, e.g. sugar is soluble in water   28, 36, 68

solution when a solute dissolves in a solvent, a solution forms   36, 38, 80, 86, 88, 89, 90, 92

sparks type of electrostatic discharge briefly producing light and sound   78, 103

species basic category of biological classification, composed of individuals that resemble one another, can breed among themselves, but cannot breed with members of another species   11, 13, 16, 18, 20, 64, 74

specific heat capacity the amount of energy needed to raise the temperature of 1 kg of a substance by 1 degree Celsius 40, 81

specific latent heat the amount of energy needed to change the state of a substance without changing its temperature; for example the energy needed to change ice at 0 °C to water at the same temperature 40

speed how fast an object travels: speed = distance ÷ time 4, 42, 43, 49, 94, 95, 96, 98, 100, 101, 106

speed–time graph a plot of how the speed of an object varies with time 95

spongy mesophyll cells found in the middle of a leaf with an irregular shape and large air spaces between them 69, 70

stable electronic structure a structure where the outer electron shell of an atom is full 86

star bright object in the sky which is lit by energy from nuclear reactions 54, 56

stem cells unspecialised body cells (found in bone marrow) that can develop into other, specialised, cells that the body needs, e.g. blood cells 63, 65

stimulant a drug that speeds up the working of the brain 8

stomata (*singular* stoma) small holes in the surface of leaves which allow gases in and out of leaves 69, 70, 71

stopping distance sum of the thinking and braking distances 96

straight line line of constant gradient 42

stratosphere a layer in the atmosphere starting at 15 km above sea level and extending to 50 km above sea level; the ozone layer is found in the stratosphere 47

subsidence settling of the ground caused, for example, by mining 38

substrate the substance upon which an enzyme acts 59

superconductors materials that conduct electricity with little or no resistance 91

sustainable development managing a resource so that it does not run out 20

S wave transverse seismic wave capable of travelling through solid but not liquid parts of the Earth 47

systolic pressure the highest point that your blood pressure reaches as the heart beats to pump blood through your body 4

T

tectonic plate a large section of the lithosphere which can move across the surface of the Earth 31

temperature a measure of the degree of hotness of a body on an arbitrary scale 9, 22, 24, 29, 35, 40, 41, 49, 51, 54, 55, 59, 60, 67, 68, 71, 73, 77, 81, 83, 91, 110

terminal speed or velocity the top speed reached when drag matches the driving force 100

therapy treatment of a medical problem 65, 109

thermal decomposition the breaking down of a compound into two or more products on heating 32, 90

thermogram a picture showing differences in surface temperature of a body 40

thinking distance distance travelled while the driver reacts before braking 96

thyroid gland gland at the base of the neck which makes the hormone thyroxin 109

tissue culture process that uses small sections of tissue to clone plants 65

total internal reflection the reflection of light inside an optically denser material at its boundary with an optically less dense material (usually air) 43

toxic a toxic substance is one which is poisonous, e.g. toxic waste 23, 26, 34

toxin a poisonous substance 6

tracer a radioactive, or radiation-emitting, substance used in a nuclear medicine scan or other research where movement of a particular chemical is to be followed 108, 109

transect line across an area to sample organisms 67

transformer device by which alternating current of one voltage is changed to another voltage 52

transition element an element in the middle section of the periodic table, between the group 1 and 2 block and the group 3 to group 0 block 90

transmitter a device which gives out some form of energy or signal, usually used to mean a radio transmitter which broadcasts radio signals 44, 46

transpiration process by which moisture is carried through plants from roots to small pores on the underside of leaves, where it changes to vapour and is released into the atmosphere 71

transverse (wave) wave in which the vibrations are at right angles to the direction in which the wave travels 42, 47, 106

trials tests to find if something works and is safe 6, 72

trophic level the stages in a food chain 14

tsunami huge waves caused by earthquakes – can be very destructive 47, 55

tumour abnormal mass of tissue that is often cancerous 106

turbine device for generating electricity – the turbine moves through a magnetic field and electricity is generated 49, 50, 110

turgor pressure the pressure exerted on the cell membrane by the cell wall when the cell is fully inflated 70, 71

U

ultrasound high-pitched sounds which are too high for detection by human ears 106

ultraviolet radiation electromagnetic waves given out by the Sun which damage human skin 47

unbalanced (forces) forces acting in opposite directions that are unequal in size 95, 96

units of alcohol measurement of alcoholic content of a drink 8

Universe the whole of space 54, 56

unstable nucleus liable to decay 110

uranium radioactive element with a very long half-life used in nuclear power stations 108, 110

V

vacuum space containing hardly any particles 41, 73, 106

variable resistor a resistor whose resistance can change 105

variation the differences between individuals (because we all have slight variations in our genes) 11, 13, 18, 61, 65

vascular bundle group of xylem and phloem cells 69, 71

vector an animal that carries a pathogen without suffering from it 6

vegan a type of diet; a person who does not eat animals or animal products 5

vegetarian a type of diet; a person who does not eat meat or fish 5

veins blood vessels that carry blood back to the heart 62, 69

velocity how fast an object is travelling in a certain direction: velocity = displacement ÷ time 95, 98, 99

voltage a measure of the energy carried by an electric current (also called the potential difference) 50, 52, 103, 104, 105

voltmeter instrument used to measure voltage or potential difference 105

volts (V) units used to measure voltage 52, 105

W

watt (W) a unit of power, 1 watt equals 1 joule of energy being transferred per second 51, 52, 97

wave oscillatory motion 42, 46, 47, 106

wavelength (λ) distance between two wave peaks 42, 45, 51, 106, 109

weight the force of gravity acting on a body 97, 100

white blood cells blood cells which defend against disease 6, 62

withdrawal symptoms reactions when a person stops taking a drug 8

work done the product of the force and distance moved in the direction of the force 97

Glossary

X

x-rays ionising electromagnetic waves used in x-ray photography (where x-rays are used to generate pictures of bones)   58, 109

xylem cells specialised for transporting water through a plant; xylem cells have thick walls, no cytoplasm and are dead, their end walls break down and they form a continuous tube   71

The key to successful revision is finding the method that suits you best. There is no right or wrong way to do it.

Before you begin, it is important to plan your revision carefully. If you have allocated enough time in advance, you can walk into the exam with confidence, knowing that you are fully prepared.

Start well before the date of the exam, not the day before!

It is worth preparing a revision timetable and trying to stick to it. Use it during the lead up to the exams and between each exam. Make sure you plan some time off too.

Different people revise in different ways and you will soon discover what works best for you.

Remember
There is a difference between *learning* and *revising*.

When you revise, you are looking again at something you have already learned. Revising is a process that helps you to remember this information more clearly.

Learning is about finding out and understanding new information.

Using the Workbook

This Workbook allows you to work at your own pace and check your answers using the detachable Answer section on pages 249–269. In addition to the exam practice questions, the Workbook also contains questions that require longer answers (Extended response questions). You will find one question that is similar to these in each section of your written exam papers. The model answers supplied for these questions give guidance about the content that should be included, but do not necessarily provide a complete response for the questions concerned.

Some general points to think about when revising

- Find a quiet and comfortable space at home where you won't be disturbed. You will find you achieve more if the room is ventilated and has plenty of light.

- Take regular breaks. Some evidence suggests that revision is most effective when tackled in 30 to 40 minute slots. If you get bogged down at any point, take a break and go back to it later when you are feeling fresh. Try not to revise when you're feeling tired. If you do feel tired, take a break.

- Use your school notes, textbook and this Revision guide.

- Spend some time working through past papers to familiarise yourself with the exam format.

- Produce your own **summaries** of each module and then look at the summaries in this Revision guide at the end of each module.

- Draw mind maps covering the key information on each topic or module.

- Review the **Grade booster checklists** on pages 242–247.

- Set up revision cards containing condensed versions of your notes.

- Prioritise your revision of topics. You may want to leave more time to revise the topics you find most difficult.

1 a A blood pressure of lower than 120/80 is normal for an adult human.

The giraffe has a much longer neck than humans.

P244 Answer

i Suggest a normal blood pressure for a giraffe. Draw a ring around your answer.

120/80   35/25   65/95   (300/200)   **[1 mark]**

ii Explain your choice *Cause longer neck and blood travel faster* **[2 marks]**

b Which units of pressure are used to measure blood pressure? Draw a ring around the correct answer.

ccMg   (mmHg)   mHg   kmGh   **[1 mark]**

c Blood pressure is written as two measurements. What are these measurements called?

Systolic pressure and *diastolic pressure* **[2 marks]**

2

Blood pressure questionnaire			
Questions	Notes	Answers Yes	No
1 Do you take regular exercise?	Strong heart muscles will lower blood pressure		✓
2 Do you eat a healthy balanced diet?	Reducing salt intake will lower blood pressure		✓
3 Are you overweight?	Being overweight by 5 kg raises blood pressure by 5 units	✓	
4 Do you regularly drink alcohol?	A high alcohol intake will damage liver and kidneys	✓	
5 Are you under stress?	Relaxation will lower blood pressure	✓	

Joe's mum has her blood pressure checked by a nurse. The nurse tells her the results and she fills in a questionnaire for the nurse.

Suggest **two** changes Joe's mum should make to lower her blood pressure.

1 *Eat a balanced diet*

2 *have regular excercises* **[2 marks]**

3

Look at the information showing percentage changes in heart disease and strokes in the UK between 1994 and 2006.

a Describe the changes shown in the chart.

the charts shows that heart disease have increased and heart stroke stayed the same **[3 marks]**

b Some people have concluded that women are fitter and healthier than men. Suggest what could be done to improve confidence in this conclusion.

.... **[2 marks]**

4 a Joe is an athlete and he is very fit. However, he still catches a cold. Explain why being fit may not keep you healthy.

If he use .... **[2 marks]**

b Joe does weightlifting to improve his fitness. This involves lifting very heavy weights a number of times. Which areas of fitness is he measuring?

The musles .... **[2 marks]**

31%
5×100÷16
5/16

1 a Draw straight lines to connect the types of food to a correct source and to their correct use (as shown).

Source	Type of food	Use
meat	carbohydrates	iron for haemoglobin
sugar	protein	Vitamin C to prevent scurvy
green vegetables	vitamins	growth
fruit juice	minerals	high energy source

[2 marks]

b Name **two** other requirements (apart from those shown in **a**) that are needed for a balanced diet.

1 ....

2 .... [2 marks]

c Rebecca eats too many carbohydrates and fats and does little exercise. She could become obese.

i What is mean by the term 'obese'?

.... [1 mark]

ii What higher risks will she have if she becomes obese?

....

.... [2 marks]

---

2 Look at the chart comparing mycoprotein and beef.

Rebecca says that mycoprotein is better than beef. Which facts **a** support and **b** disagree with Rebecca's conclusion?

Chart data:
- protein (g per 100 g): beef 20.0, mycoprotein 13.9
- fat (g per 100 g): beef 4.6, mycoprotein 4.0
- dietary fibre (g per 100 g): beef 0, mycoprotein 7.0
- cholesterol (mg per 100 g): beef 59, mycoprotein 0
- energy (kJ per 100 g): beef 515, mycoprotein 380

a ....

b .... [2 marks]

---

3 a Simon has a mass of 40 000 g. Calculate his estimated average daily requirement (EAR) for protein in grams. Use this formula: EAR in g = 0.6 × body mass in kg. Show your working.

EAR = ... g [2 marks]

b Simon has a sister called Karen. He is worried that she is not eating enough food. Suggest why Karen may have chosen not to eat enough food to meet her daily requirements.

....

....

.... [2 marks]

---

4 Glycogen is a large molecule made up of hundreds of glucose molecules joined together.

glucose units

glycogen is a complex carbohydrate

a What group of food chemicals does glycogen belong to?

.... [1 mark]

b Proteins are also made of hundreds of small molecules that are joined together in a chain. Write down the name of these molecules.

.... [1 mark]

1 a Draw a line from each type of pathogen to a correct example.

Type of pathogen

| fungi |
| bacteria |
| viruses |
| protozoa |

Example

| cholera |
| athlete's foot |
| lung cancer |
| malaria |
| flu |

[4 marks]

b Red–green colour blindness is not contagious. Explain why.

.... [2 marks]

2 New drugs are continually being developed. Why are new drugs tested before they can be used?

.... [2 marks]

3 Complete the sentences about infectious diseases. Use words from this list:

antibodies   antibiotics   antigens   hormones   pathogens   toxins   vectors

a The symptoms of an infectious disease are caused by ....

They produce chemical waste, which contains ....

b Each disease-causing microorganism has its own .... so

the body needs specific .... to combat them.   [2 marks]

4

adult emerges

female mosquito
feeds on blood

eggs

surface
of water

larva → developing larva

developing larva

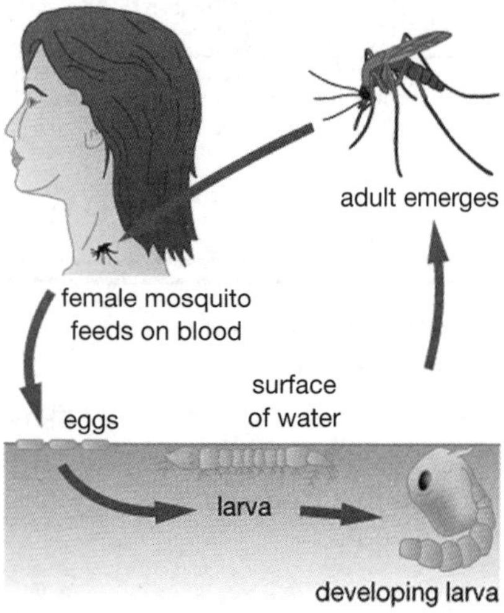

Look at the diagram. It shows how mosquitoes act as a vector.

a What is meant by a vector?

b Using the diagram, name a parasite and its host.

....

.... [1 mark]

5 a The graph shows the levels of immunity to a disease using passive and active immunity methods. Describe the difference in immunity:

Increasing level of immunity

—— Passive immunity
--- Active immunity

0 10 20 30 40 50 60 70 80 90 100
Days after treatment

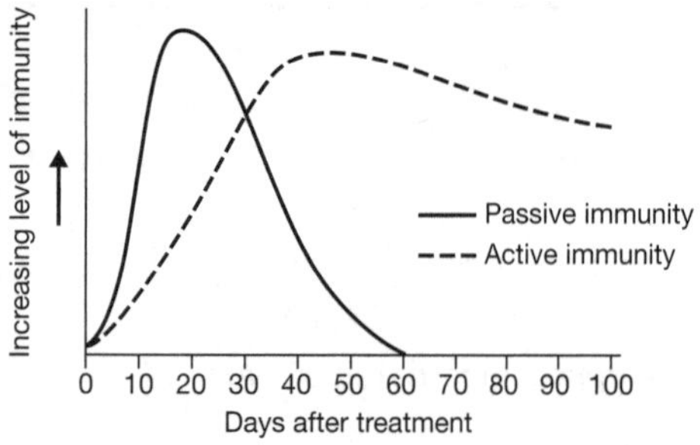

i at 20 days .... [1 mark]

ii after 60 days .... [1 mark]

b Explain why there is a difference in immunity between passive and active immunity.

....

....

.... [4 marks]

1 Complete the table showing information about senses.

D–C

Sense organ (receptor)	Sense	Stimulus
skin	touch	pressure, temperature
	taste	chemicals in food
nose	smell	
eyes	sight	light
	hearing and balance	sound

[3 marks]

2 Rashid does his homework on the eye and how it works.

G–E

a Complete the diagram by writing the correct labels in the three boxes. [3 marks]

b There are **three** mistakes in Rashid's homework. This is what he writes.

The light rays are reflected by the cornea and lens. An image is formed on the optic nerve. Nerve impulses are then sent to the spinal cord. The amount of light entering the eye is controlled by the iris.

D–C

Write down the incorrect word(s) and their correct replacements.

.... should be ....

.... should be ....

.... should be .... [3 marks]

c Write down **two** ways in which vision will be affected if only one eye is used to look at an object.

1 ....　　2 .... [2 marks]

3 Complete the sentences by putting a ring around the correct answers.

G–E

The nerve impulse is:　　a chemical code　　an electrical impulse　　a physical push.

The nervous system consists of the peripheral nervous system and the:　　NS　　ATP　　CNS　　[2 marks]

4

Look at the diagram of a motor neurone.

a Which part carries the nerve impulse?

.... [1 mark]

D–C

b Look at the sequences A, B, C and D in a reflex arc. Only one sequence is correct.

A　effector⇒sensory neurone⇒motor neurone⇒central nervous system⇒response

B　stimulus⇒receptor⇒sensory neurone⇒motor neurone⇒central nervous system⇒response

C　stimulus⇒receptor⇒sensory neurone⇒central nervous system⇒motor neurone⇒response

D　response⇒receptor⇒sensory neurone⇒central nervous system⇒motor neurone⇒stimulus

Which sequence is correct? .... [1 mark]

1 a Complete the sentences about smoking. Choose from these words:

addicted   infectious diseases   rehabilitation   withdrawal symptoms

Rosie is finding it difficult to give up smoking because she is

.... to cigarettes.

As she gives up smoking she will suffer from ....   **[2 marks]**

b Which component in cigarette smoke:

i is an irritant? ....

ii is addictive? ....   **[2 marks]**

c Explain what causes a smoker's cough.

....

....

....   **[4 marks]**

d Look at the graph. It shows death rates in men from lung cancer in the 1950s, in relation to the type of tobacco they smoked.

i Work out the death rate per 100 000 men in the UK for all types of smoking.

....   **[1 mark]**

ii The present death rate from lung cancer in the UK is about 60 deaths per 100 000. Explain the reasons for the change in death rate.

....

....

....   **[2 marks]**

2 Draw a straight line from each type of drug to a correct example.

Types of drug	Example
depressant	LSD
painkiller	alcohol
stimulant	paracetamol
hallucinogen	caffeine

[3 marks]

....

3 Explain why there is a legal limit for the level of alcohol in the blood/breath of drivers.

....

....

....

....   **[4 marks]**

1 a i What is the normal core temperature of the body? .... [1 mark]

ii Name **two** ways of measuring body temperature.

1 ....   2 .... [2 marks]

b Complete the sentences with words from this list:

> **sweating a lot**   **exercising**   **respiring**   **wearing few clothes**   **less sweating**

Heat is gained/retained by: ....

Heat is lost by: .... [2 marks]

G–E

c If the body gets too hot you can suffer from dehydration. Explain why.

.... [2 marks]

d What is meant by homeostasis?

.... [1 mark]

D–C

2 a Complete the sentences about diabetes. Use words from this list:

> **blood**   **insulin**   **faster**   **liver**   **neurones**   **pancreas**   **same as**   **slower**

The level of glucose in the blood is controlled by the hormone called

.... The hormone is produced in the ....

This hormone is carried in the .... so its effects

are .... than responses controlled by the nervous system. [4 marks]

G–E

b Type 1 diabetes is treated by hormone injections. Explain why Type 2 diabetes can be controlled by diet.

....

.... [3 marks]

D–C

c Look at the graph showing the blood sugar levels of two people after drinking a glucose solution.

i Why did the blood glucose levels rise in the first hour?

....

....

....

.... [1 mark]

ii Explain why the results for the person with diabetes are different from the person who doesn't have diabetes.

....

....

....

....

....

....

.... [2 marks]

G–E

1 What advantages does a plant shoot gain by growing towards the light?

Put a tick (✓) in the boxes next to the **two** correct answers.

They get more light. ☐

They get more water. ☐

They get more oxygen. ☐

They make more glucose. ☐

They respire faster. ☐ [2 marks]

2 Describe an experiment to show that shoots grow towards the light. Write one line each under the following headings:

How the experiment was set up.

....

The control (comparison) experiment used.

....

Results

....

Conclusion

.... [5 marks]

3 Charlotte grows fruit trees.

a She takes cuttings of the best trees and dips the ends in rooting powder.

Describe and explain the effect the rooting powder has on the cuttings.

....

.... [2 marks]

b Charlotte has many weeds in her lawn. She uses a selective weedkiller.

i Explain what is meant by a selective weedkiller.

....

.... [2 marks]

ii How does a selective weedkiller work?

....

.... [2 marks]

4 Complete the sentences about tropisms. Use words from this list:

auxins   antigens   geotropic   gravity   heat   light   negatively   positively

Since shoots grow towards the light they are called ....
phototropic.

Since roots grow with the pull of .... they are called

positively ....

These reactions involve plant hormones called .... [4 marks]

1 Which human characteristics are caused by both genetic and environmental factors?

Put a tick (✓) in the boxes next to the **three** correct answers.

body mass	☐	intelligence	☐
broken teeth	☐	lobed ears	☐
height	☐	scars	☐

[3 marks]

G–E

---

2 Complete the sentences by drawing a ring around the correct answer.

Most body cells contain chromosomes in: genes   triplets   matching pairs.

An example of an inherited disorder is: sickle cell anaemia   malaria   measles   [2 marks]

G–E

---

3 Complete the sentences using words from this list.

cytoplasm   **environmental**   **enzymes**   **genes**   **genetic**   **nucleus**

Chromosomes are found in the .... of cells.

They carry information in the form of .... which control

characteristics caused by .... factors.   [3 marks]

G–E

---

4 Mutations can cause genetic variation in organisms.

a What is a mutation? ....   [1 mark]

b Explain **two** other causes of genetic variation.

1 ....

2 ....   [4 marks]

D–C

---

5 Look at the diagram showing chromosome numbers. Complete it by writing the correct number of chromosomes in the empty boxes.

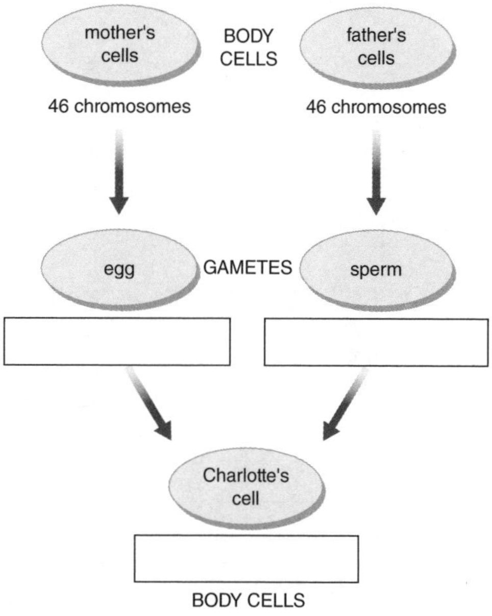

mother's cells   BODY CELLS   father's cells

46 chromosomes   46 chromosomes

egg   GAMETES   sperm

☐   ☐

Charlotte's cell

☐

BODY CELLS   [3 marks]

G–E

The blue whale is a mammal. It migrates long distances through different oceans.

The blue whale controls its body temperature in a similar way to humans. It also has a similar body temperature to humans.

The diagrams show sections through whale skin. Use information in the diagrams to explain how and why the blue whale controls its body temperature.

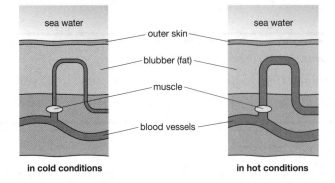

❗ The quality of written communication will be assessed in your answer to this question.

....
....
....
....
....
....
....
....
....
....
....
....
....
....
....
....
....
....
....
....
....
.... [6 marks]

1 Lions are members of the animal kingdom.

a What characteristics do all animals share?

.... [1 mark]

G-E

b This table shows the classification of lions.
Complete it by inserting the missing groups.

kingdom	Animal
	Chordate
	Mammal
	Carnivore
family	Felidae
	Panthera
	leo

[5 marks]

D-C

2 a There are seven different types of lion alive today.

Complete the sentences about lions using words from this list.

    all   **different**   **habitat**   **kingdom**   **similar**   **variation**

All lions share .... characteristics.

If they live in a similar ...., the lions are more alike.

Members of one type of lion share more characteristics with each other than with other
types of lion.

Even in the same group there are differences. This is called .... [3 marks]

G-E

b What is meant by the term **species**?

.... [2 marks]

c Lions and tigers belong to the same family of cats.
Look at the table. It shows the Latin names of
some different cats.

Common name	Latin name
bobcat	*Felix rufus*
cheetah	*Acinonyx jubatus*
lion	*Panthera leo*
ocelot	*Felix pardalis*

i Two of the cats are more closely related.
Write down the common names of these
two cats.

.... [1 mark]

ii Explain your answer to part (ci).

.... [1 mark]

D-C

3 Scientists discovered a fossil of an extinct animal called *Archaeopteryx*. It had some bird features
and some reptile features. Explain why there are always going to be some organisms like this that
are difficult to classify.

....
.... [2 marks]

D-C

1 a Grass belongs to the first trophic level and is a producer.

```
          ┌──────────┐        ┌──────────┐      ┌──────────┐
          │ beetles  │───────▶│   mice   │─────▶│   owls   │
          └──────────┘        └──────────┘      └──────────┘
┌──────────┐    │   │
│  grass   │────┘   │
└──────────┘        ▼
          ┌──────────┐
          │ hedgehogs│──────────────────────────────┘
          └──────────┘
```

G–E

i What is a trophic level?

.... [1 mark]

ii Grass is a green plant. Write down **one** other type of organism that can be a producer.

.... [1 mark]

iii The mice in this habitat were killed by poison.

Suggest what might happen to the numbers of hedgehogs. Explain your answer.

.... [2 marks]

b It is possible to construct a pyramid of biomass for this food chain.

i What does a pyramid of biomass show?

....
.... [2 marks]

D–C

ii Explain why the pyramid of biomass for this food chain would be a different shape than a pyramid of numbers.

....
.... [2 marks]

2 Look at the diagram. It shows the energy transfer from crops to a cow.

G–E

a What is the original source of energy for this food chain?

1022 kJ in heat loss

Sun

1909 kJ in waste

3056 kJ energy in 1 m²

b The cow loses 1022 kJ in heat loss.

Write down the process occurring in the cow that generates this heat.

.... [1 mark]

D–C

c Farmers often try to reduce the energy lost from cows.

i Suggest one way that they can do this.

.... [1 mark]

ii Suggest why they want to do this.

.... [1 mark]

2 Finish the sentences about rotting apples.

Choose the best words from this list:

    decay   **digestion**   **nitrogen**   **oxygen**   **preserved**   **recycled**

The apples are starting to break down. This is called ....

They are breaking down to release important elements such as carbon and

....

These elements are .... and used by living plants and animals.   **[3 marks]**

G–E

2 The diagram shows the carbon cycle.

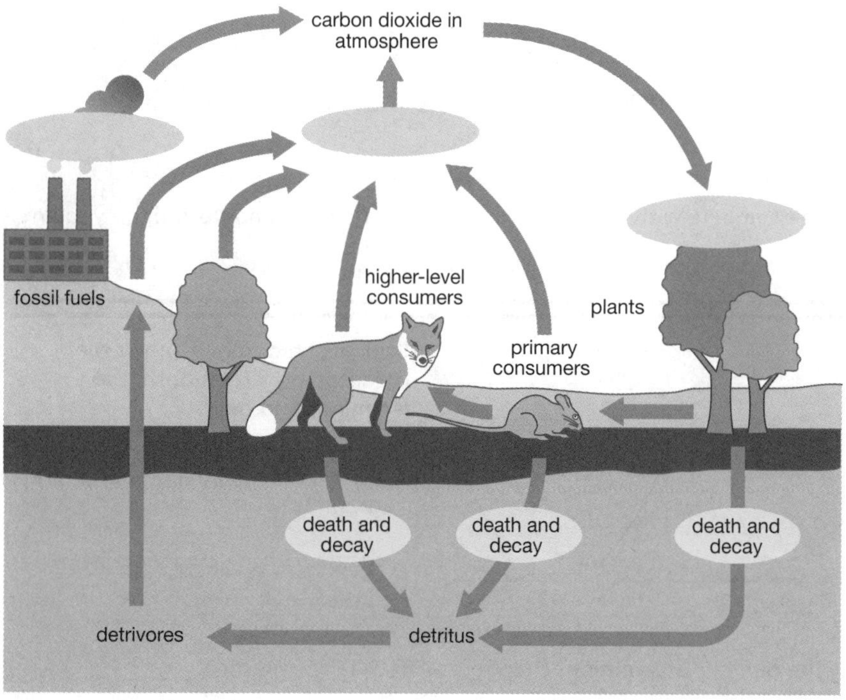

a Finish labelling the cycle by writing the correct processes in the three blank labels.   **[3 marks]**

b Decomposers return carbon dioxide to the air.

They will release carbon dioxide more slowly from waterlogged soils.

Explain why.

....

....   **[2 marks]**

D–C

3 a What percentage of the air is nitrogen gas?

Put a ring around the answer in this list:

    0.03%   **3%**   **20%**   **50%**   **78%**

b Plants cannot take up nitrogen directly from the air. Why is this?

....   **[1 mark]**

G–E

c Why do plants need nitrates?

....   **[1 mark]**

D–C

1 Read the information about the red and grey squirrels.

There are two main species of squirrel living in Britain, the red squirrel and the grey squirrel. The grey was introduced from America. The number of reds has gone down in Britain over the last 60 years. They are now rare. There have been a number of studies to try to find out why the number of reds is declining.

In coniferous woodland the reds are lighter animals and so spend more time up in the trees, feeding on the seeds from pine cones.

It is in deciduous woodland that the reds are disappearing. Both types try to feed on acorns on the floor of the forest but the greys can digest the acorns more easily.

a Some people thought that the grey squirrel might have introduced a disease into the country. What evidence is there that the grey squirrel did not pass on a disease to the red squirrels?

.... [1 mark]

b What do red and grey squirrels compete for?

.... [1 mark]

c What do red squirrels compete with each other for that they do not compete with grey squirrels for?

.... [1 mark]

2

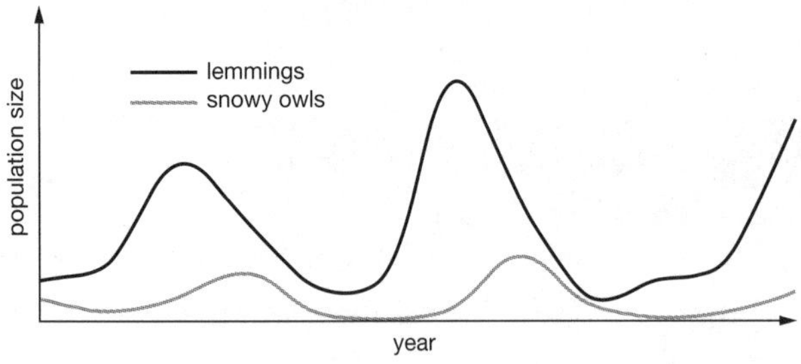

population size / year — lemmings, snowy owls

Look at the graph. It shows the predator–prey relationship of lemmings and owls.

a Why is this called a predator–prey graph?

....

.... [2 marks]

b Write about why the number of lemmings change, as shown in the graph.

....

.... [2 marks]

3 Oxpeckers are birds that often live on buffalo. Buffalo often suffer from insect parasites. The oxpeckers and the buffalo both benefit from the relationship.

a Describe how they both benefit.

.... [2 marks]

b Write down the name given to this type of relationship.

.... [1 mark]

4 Pea plants contain bacteria in special nodules on their roots. Pea plants are pollinated by bees. These bees often have small animals called mites that feed on fluid from the bee's body.

a Write down an organism that is acting as a parasite in these feeding relationships.

.... [1 mark]

b The pea plant and the bee both gain from their relationship. Explain how.

.... [2 marks]

1 a Polar bears are adapted to hunt.

Put a tick (✓) in the box that shows how polar bears are adapted to hunt.

Eyes on the front of their head	
Fat under the skin	
Streamlined body	
Thick fur	

[1 mark]

b A chameleon is a type of lizard that changes its colour according to its environment. It feeds on insects, such as grasshoppers. It is fed on by snakes.

The chameleon can move both its eyes so that they can point to the side or point directly forward.

i Explain how the movement of the eyes can help it to catch insects.

....
.... [2 marks]

ii Explain how the movement of the eyes can help it avoid being eaten by snakes.

....
.... [2 marks]

iii Suggest how the usual colour of the chameleon (green) helps it to avoid being eaten by snakes.

....
.... [2 marks]

G-E

2 a Polar bears are adapted to live in the cold. Suggest ways that the polar bear may be adapted to live in the cold.

....
....
.... [3 marks]

b The black bear and the polar bear both live in Canada. Both types of bears spend the winter months buried in holes or dens.

Explain why they do this.

....
....
.... [3 marks]

D-C

3 a Cacti live in deserts.

The cactus has spines instead of leaves. Explain why.

....
.... [2 marks]

b Lizards called gila monsters live in the same habitat as cacti. They hunt small mammals.

In the morning they lie out in the Sun for some time before they hunt. Explain why.

....
.... [2 marks]

D-C

1 **a** The spines on a holly leaf help protect the leaves. If there was a change in the climate, less plant food may be available for animals.

If this happened, which holly plants are more likely to survive and why?

....

.... [2 marks]

b A change in the climate could lead to a change in the population of holly bushes over a long period of time. What is such a change called?

.... [1 mark]

c What is likely to happen to organisms that cannot change if the environment changes?

.... [1 mark]

2 Question 1 on page 146 includes a passage about competition between the red and grey squirrel. Re-read the passage before attempting these questions.

a What adaptation allows the grey squirrels to be better adapted to living in deciduous woodland?

.... [1 mark]

b Many scientists hope that a new strain of the red squirrel will evolve that can digest acorns. Use Darwin's theory of natural selection to explain how this might come about.

....

....

....

.... [4 marks]

3 The following article gives information on the superbug MSRA. Read it carefully.

Where did it come from?

MRSA evolved because of natural selection. There are lots of different strains of the bacteria. Each strain has slightly different DNA. The DNA is also constantly mutating as the bacteria reproduce. Some of these mutations will be more resistant to antibiotics than others. When people take antibiotics, the less resistant strains die first. The more resistant strains are harder to destroy. If people stop taking the antibiotics too soon, the resistant strains survive.

Use Darwin's theory of natural selection to explain how MRSA has evolved.

....

.... [2 marks]

4 On his voyage on the *Beagle*, Charles Darwin visited many small islands. He made a number of observations that helped him to develop his theory of natural selection. One observation was that on small islands, animals often evolve to produce smaller animals than on the mainland.

When Darwin returned from his voyage he was rather worried about publishing his ideas about natural selection. Suggest why that was.

....

....

.... [2 marks]

1 a Human activity is causing an increase in the levels of carbon dioxide in the air.

i How is human activity causing the release of carbon dioxide?

.... **[1 mark]**

ii Why is the release of carbon dioxide increasing?

.... **[1 mark]**

iii Suggest **two** effects this increase may have on the environment.

....

.... **[2 marks]**

b The ozone layer in the Earth's atmosphere protects us from harmful ultraviolet rays. Chemicals are destroying the ozone layer.

i Write down the name of the chemicals. .... **[1 mark]**

ii The overuse of these chemicals has caused an increase in skin cancer. Suggest a reason why.

.... **[1 mark]**

---

2 Look at the graph. It shows the past, present and predicted future world human population.

a The rapid increase in the world's population means that certain resources are running out.

What word is used to describe a resource that is in limited supply?

Underline the answer in this list:

> **extinct**   **finite**   **infinite**   **polluting**   **recycled**

b Write down **one** resource used by people that is in limited supply. .... **[1 mark]**

c Human population is in the rapid growth stage. Write down the name given to this stage of growth. **[1 mark]**

---

3 Scientists can look at the variety of animal species living in a stream when they want to measure how polluted the water is.

a Why can animals be used as an indicator of pollution?

.... **[1 mark]**

b Write down the name given to species that are used to measure levels of water pollution.

.... **[1 mark]**

c

Animal	Sensitivity to pollution
stonefly larva	sensitive
water snipe fly	sensitive
alderfly	sensitive
mayfly larva	semi-sensitive
freshwater mussel	semi-sensitive
damselfly larva	semi-sensitive
bloodworm	tolerates pollution
rat-tailed maggot	tolerates pollution
sludgeworm	tolerates pollution

Look at the table. It shows the sensitivity of different animals to pollution.

A river sample contained mussels, damsel fly larvae and bloodworms, but no mayfly or stonefly larvae. Use this information to explain how you can tell that the river is polluted.

....

.... **[2 marks]**

1 a The dodo was a large flightless bird. It no longer exists.

i Write down the term used to describe an animal or plant that no longer exists.

.... [1 mark]

ii Suggest **two** reasons why the dodo no longer exists.

....

.... [2 marks]

G-E

b The giant panda is an endangered species. Pandas live in bamboo forests in a remote part of China. Their forests habitats are being destroyed.

i Write down **two** ways that scientists can help protect pandas.

....

.... [2 marks]

ii Suggest **two** reasons why saving the panda might help the people who live in the same habitat.

D-C

....

.... [2 marks]

2 Some countries want to hunt whales for food.

a Suggest **one** argument for and **one** argument against hunting whales.

....

....

....

D-C

.... [2 marks]

b It is very difficult to stop people hunting whales.

Suggest **one** reason why.

.... [1 mark]

3 If treated properly, fish can be a sustainable resource.

G-E

a What is meant by a sustainable resource?

.... [1 mark]

b To try and achieve this, the government has set fish quotas.
Fishermen can only catch a set amount of fish at any one time.
Also, the size of the individual fish that they catch has to be above a certain level.
These regulations should help maintain the population of fish in the sea.

Explain why.

D-C

....

....

.... [2 marks]

B2 Understanding our environment

The diagram shows the flow of energy through a food chain.

```
                          → 9000 units            → 800 units
                        ┌─┘                      ┌─┘
sunlight →  ┌──────────┐     ┌──────────┐     ┌──────────┐
            │  wheat   │  →  │   cows   │  →  │  humans  │
            │10 000 units│     │1000 units│     │200 units │
            └──────────┘     └──────────┘     └──────────┘
```

Explain how energy passes through this food chain and explain the pattern of figures.

❶ The quality of written communication will be assessed in your answer to this question.

....

....

....

....

....

....

....

....

....

....

....

....

....

....

....

....

....

....

....

....

....

....

....

....

....

.... [6 marks]

1 a Write three examples of fossil fuels.

gas, oil and coal ✓ [1 mark]

b Fossil fuels are non-renewable. What is meant by this?

Fossil fuels that cannot be use to make a new [1 mark] material.

2 a Crude oil is separated by fractional distillation into fractions. Why does this work?

Because different fractions have different [1 mark] boiling point,

b Liquid petroleum gas (LPG) is one fraction. Which two gases does LPG contain?

Propane and butane [1 mark]

c Write down four other fractions that can be made during the fractional distillation of crude oil.

LPG, petrol, desel and bitumen ✓ [2 marks]

d Crude oil is a mixture of hydrocarbons which is separated by fractional distillation.

i Label the diagram A where the crude oil is heated. [1 mark]

ii Label the diagram B where the fraction bitumen 'exits' from. [1 mark]

iii Label the diagram C at the coldest part. [1 mark]

iv Which fraction 'exits' from the coldest part?

The fraction that is at the top. It is LPG. [1 mark]

3 When crude oil is transported by sea, accidents can happen and cause damage. Explain how this damage affects the natural world.

Water in the sea get greasy so bird se can't drink water and it Pollute [2 marks] sea water.

4

a Paraffin can be broken down or 'cracked'. Label the diagram of the apparatus needed. [2 marks]

b What conditions are needed to crack liquid paraffin?

By heating it at high [2 marks] temprature. + catalys

c Cracking breaks down long-chain molecules called alkanes. When a large alkane is cracked it becomes a smaller alkane and an alkene. Explain why an alkene is a different type of hydrocarbon to an alkane.

Alkenes has less hydrogens and double bonds (=). ✓ [1 mark]

d What are alkenes useful for making? Plastics. ✓ [1 mark]

1 Look at the table.

a One of the key factors when choosing the best fuel for a given use is missing. Which one?

.... [1 mark]

Characteristic	Coal	Petrol
energy value	high	high
availability	good	good
storage	bulky and dirty	volatile
toxicity	produces acid fumes	produces less acid fumes
pollution caused	acid rain, carbon dioxide, soot	carbon dioxide, nitrous oxides
ease of use	easier to store for power stations	flows easily around engines

G–E

b Use the information to compare how coal and petrol are different in terms of storage and toxicity.

.... [2 marks]

D–C

2 a Why are fuels burnt, and what is needed so that they can burn?

....

.... [2 marks]

b What are the two products that are made in the complete combustion of a hydrocarbon fuel?

....

.... [2 marks]

G–E

c An experiment can show that two products are made in the complete combustion of a hydrocarbon fuel.

i Write down a **word equation** for the complete combustion of a hydrocarbon fuel.

.... [2 marks]

D–C

ii Describe an experiment to show what the products of combustion are when a candle burns in a plentiful supply of air.

.... [2 marks]

3 a i Why does incomplete combustion sometimes take place?

.... [1 mark]

ii What products are made?

.... [2 marks]

G–E

b Complete combustion is better than incomplete combustion. Explain why.

....

....

.... [3 marks]

D–C

1 a One gas that is found in clean air varies in proportion considerably. Which one?

.... **[1 mark]**

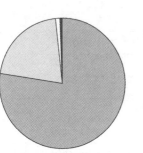

b Label the diagram with the percentages of nitrogen, oxygen and carbon dioxide found in clean air. (Ignore noble gases). **[3 marks]**

c Mark in the percentages of the gases. **[3 marks]**

2 a How do photosynthesis, respiration and combustion affect the levels of carbon dioxide and oxygen in the air?

	Photosynthesis	Respiration	Combustion
carbon dioxide			
oxygen			

[3 marks]

b The percentages of the three main gases found in clean air do not change very much because there is a balance between three of the processes that use or make carbon dioxide and use or make oxygen. Explain how the balance is maintained.

....

....

.... **[6 marks]**

3 a Look at the table. Sulfur dioxide causes acid rain. List three problems that acid rain can cause.

Pollutant	Environmental effects
carbon monoxide	Poisonous gas formed by incomplete combustion of petrol – or diesel-powered motor vehicles
oxides of nitrogen	Photochemical smog and acid rain, formed by reaction of nitrogen and oxygen at very high temperatures such as those in an internal combustion engine
sulfur dioxide	Acid rain formed from sulfur impurities when fossil fuels burn erodes stonework, corrodes metals, kills plants and fish, and can lead to trees dying

....

....

.... **[3 marks]**

b It is important to control levels of atmospheric pollution. Use the table to explain why pollution should be controlled from petrol-powered cars.

....

....

.... **[3 marks]**

1 a Which two elements are chemically combined to make a hydrocarbon?

.... [1 mark]

b Which of these compounds are hydrocarbons? Put a ring around the correct answers.

A

B

C CH_4

D

[1 mark]

c Look at the formula for propane.

What kind of compound is it?

.... [1 mark]

d Butanol C_4H_9OH is **not** a hydrocarbon.

Explain why.

.... [1 mark]

e Butene is an alkene.

Explain how you know.

.... [1 mark]

f Bromine water is used to test for an alkene. What do you see if an alkene is tested?

....

.... [2 marks]

2 a What is a polymerisation reaction?

....

.... [2 marks]

b Which polymer is made from propene?

.... [1 mark]

c Which monomer makes the polymer poly(styrene)?

.... [1 mark]

d Which molecule is a polymer? Put a ring around A, B, C or D.

A

B

C

D

[1 mark]

e Write down **two** conditions needed for polymerisation.

....

.... [2 marks]

G–E

D–C

G–E

D–C

1 Look at the table to answer these questions.

	Poly(ethene)	Poly(styrene)	Nylon
Can be made into a fibre	no	no	yes
Waterproof	yes	yes	yes
Rigid	no	yes	no
Heat insulator	no	yes	no

a Nylon is the best of these three materials to make a raincoat with. Explain why.

....

.... [1 mark]

b Poly(styrene) is the best of these three materials to make a cup for holding hot drinks. Explain why.

....

.... [1 mark]

c Poly(ethene) is often used to make plastic bags. Explain why you cannot tell it is the best material of these three, using only the information in the table.

....

.... [1 mark]

d Polymers are better than other materials for some uses. Give one example of the use of a polymer and explain why that polymer was suitable for the use.

....

....

.... [2 marks]

e Nylon has a disadvantage when used for outdoor clothing. What is it?

.... [1 mark]

2 a Many polymers are non-biodegradable. Explain what this means.

....

.... [2 marks]

b Disposing of non-biodegradable polymers causes problems. Explain the problems for each of the disposal methods listed below.

i Landfill sites .... [1 mark]

ii Burning waste plastics .... [1 mark]

iii Recycling .... [1 mark]

c Scientists are developing new types of polymers that are more easily disposed of after use. Explain what is different about the ways that these plastics can be disposed of.

....

....

.... [2 marks]

1 a Explain why cooking is a chemical change. Use each of these words.

  substance   irreversible   energy   reversed

  ....

  ....

  .... [2 marks]

G–E

  b What happens to the protein molecules in eggs and meat when eggs and meat are cooked and what is this process called?

  ....

  ....

  .... [2 marks]

D–C

2 Match the type of food additive to its function.

antioxidants		help mix oil and water
food colours		improve the taste
flavour enhancers		prevent reaction with oxygen
emulsifiers		improve the appearance

[2 marks]

G–E

3 a How does baking powder help to make cakes rise?

  ....

  .... [1 mark]

  b Limewater is used to test for a substance. Write down the substance and what you will see.

delivery tube

limewater

  ....

  .... [2 marks]

G–E

  c When sodium hydrogencarbonate (baking powder) is heated it decomposes.

  Write down the word equation for this reaction.

  .... [1 mark]

  d Write down the balanced symbol equation for the decomposition of sodium hydrogencarbonate to make sodium carbonate, carbon dioxide and water. You may look up the formulae for sodium hydrogencarbonate and sodium carbonate.

  .... [2 marks]

D–C

1 a Write down one source of natural perfume.

.... [1 mark]

b If a perfume is not from a natural source it is:

.... [1 mark]

c To make a perfume, a type of chemical is made by mixing alcohol with an acid.

i Write down a word equation for the reaction to make this type of chemical.

.... [1 mark]

ii Describe how you would make a sample of this chemical in a school laboratory. Draw a diagram of the apparatus you would use.

....

....

....

....

....

....

....

[3 marks]

2 A good perfume needs to have several properties. These are listed in the boxes. Draw a straight line to match the best reason to the property needed.

evaporates easily	it can be put directly on the skin
non-toxic	it does not react with perspiration
insoluble in water	it does not poison people
does not irritate the skin	it cannot be washed off easily
does not react with water	its particles can reach the nose

[4 marks]

3 Complete these sentences by choosing the best words from the list.

insoluble   soluble   solute   solution   solvent

Water cannot be used to remove varnish from nails, water is not a ....

for nail varnish. Nail varnish is .... in water. Nail varnish is

.... in nail varnish remover. A solvent dissolves a

.... to make a .... [3 marks]

4 a A solute and a solvent that do not separate is a .... [1 mark]

b Give two uses of esters.

.... [1 mark]

5 Explain why new cosmetic products need to be thoroughly tested before they are permitted to be used.

....

.... [2 marks]

1 a Match the ingredients of paint to their function.

solvent	substance that gives the paint its colour
binding medium	thins the paint making it easier to spread
pigment	sticks the pigment in the paint to the surface

[2 marks]

b What is an oil paint?

.... [1 mark]

c Paint is used for different purposes. Write down why paint would be used in these situations.

i on a front door

.... [1 mark]

ii in a living room

.... [1 mark]

iii on a car

.... [1 mark]

iv in a painting

.... [1 mark]

d Paint is a **colloid**. Explain the meaning of the term 'colloid'.

.... [1 mark]

e Emulsion paint is a water-based paint. It is made of tiny droplets of a liquid in water, which is called an emulsion. When emulsion paint has been painted on to a surface as a thin layer, it is left to dry. What happens as it dries?

.... [1 mark]

2 a What do thermochromic pigments do?

.... [1 mark]

b Give one use for a thermochromic pigment.

.... [1 mark]

c A thermochromic pigment changes colour at 45 °C. Suggest two objects that the pigment could be used in, explaining your reasons.

....
.... [2 marks]

3 Why do phosphorescent pigments glow in the dark?

....
.... [2 marks]

C1 Carbon chemistry 159

Perfumes and nail varnish are cosmetics. Synthetic perfumes are made from esters. Perfume needs to have a number of properties, including having a pleasant smell and being insoluble in water.

Describe a simple experiment to make an ester. List three properties a perfume must have (other than those listed above), giving reasons why it must have these properties.

Explain what you would use to remove nail varnish and why you would not use water.

❗ The quality of written communication will be assessed in your answer to this question.

....
....
....
....
....
....
....
....
....
....
....
....
....
....
....
....
....
....
....
....
....
....
....
....
....
....
....
....
....
....

[6 marks]

1 a Label the diagram of the structure of the Earth.

[3 marks]

b The outer part of the Earth is called the lithosphere. What makes up the lithosphere?

....

.... [2 marks]

c Describe why it is difficult to study the structure of the Earth.

....

.... [2 marks]

2 a Describe the speed of the movement of tectonic plates and, as a result, the movement of the continents.

....

.... [2 marks]

b What other things happen when tectonic plates move?

....

.... [2 marks]

c Suggest why most scientists now accept plate tectonic theory.

....

.... [2 marks]

3 a Describe how the size of crystals changes when molten rock cools down in different ways.

....

....

.... [4 marks]

b Describe the two ways that volcanoes can erupt.

....

.... [2 marks]

c Explain why some people choose to live near volcanoes.

.... [1 mark]

d Suggest two reasons why geologists study volcanoes.

....

.... [2 marks]

G–E

D–C

G–E

D–C

G–E

D–C

1 a Write down three rocks that are used in the construction of buildings and roads.

....

....

.... **[3 marks]**

b Explain what environmental problems occur when rocks are quarried or mined from the ground.

....

.... **[2 marks]**

c Put these materials into order of hardness.

      granite   **limestone**   **marble**

Least hard ....

.... Hardest   **[1 mark]**

d Aluminium, brick and glass are all manufactured. Finish the table to show the raw materials used to make each one.

building material	aluminium	brick	glass
raw material			

[3 marks]

e Cement is used as a building material. Describe how cement is made.

....

.... **[1 mark]**

2 a Limestone and marble are two forms of the same chemical. Which chemical?

.... **[1 mark]**

b What does limestone decompose into when it is heated?

....

.... **[2 marks]**

c Describe how concrete is made, and how it can be strengthened.

....

....

.... **[2 marks]**

d Calcium carbonate decomposes at high temperatures. Write a word equation for this reaction.

.... **[1 mark]**

C2 Chemical resources

1 a Explain how copper is extracted from its ore.

....

.... [2 marks]

b What is meant by 'reduction'?

.... [1 mark]

c Copper needs to be purified. What is the process used to purify copper (shown in the diagram below)?

.... [1 mark]

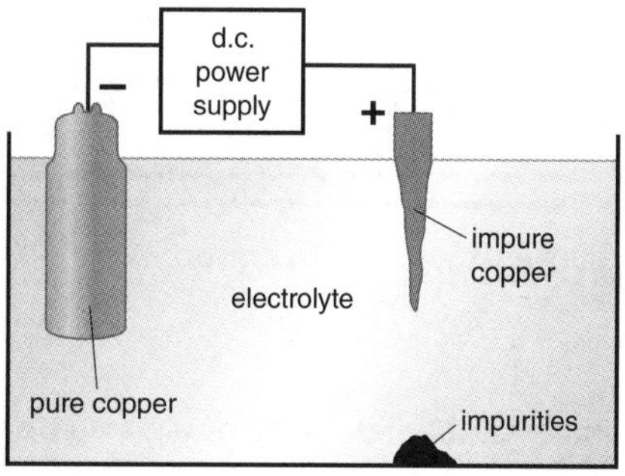

d Label the cathode on the diagram. [1 mark]

e Give **two** advantages and **two** disadvantages of recycling copper.

Advantages ....

.... [2 marks]

Disadvantages ....

.... [2 marks]

2 a Choose **three** properties from the list below that would be needed for a metal power cable carrying high-voltage electricity overhead.

waterproof	high density	good electrical conductivity	weak
low density	high tensile strength		

....

.... [3 marks]

b Most metals form alloys.

i Draw a straight line to match the real-world use to the alloy. [1 mark]

Then

ii Draw a straight line to match each alloy to the metals it is made from.

musical instruments	amalgam	contains copper and zinc
joins electrical wires	brass	contains mercury
tooth fillings	solder	contains lead and tin

[2 marks]

1 a What three things are needed for rusting to happen?

.... [1 mark]

b Why is rusting an oxidation process?

.... [1 mark]

c In winter, icy roads are treated with rock salt. Why is this a problem for car bodies made from steel?

.... [1 mark]

d Aluminium does not corrode in moist air. Explain why.

.... [1 mark]

e Write a word equation for rusting.

.... [2 marks]

2 a Steel is an alloy of iron. Give **two** advantages of steel over iron.

.... [1 mark]

b Steel and aluminium can be used to make car bodies. Write down **two** disadvantages of using aluminium.

.... [2 marks]

c Finish this table. Write down the reasons why these materials are used in a car.

material and its use	one reason for using the material
copper in electrical wires	
plastic bumpers	
PVC wire covering	

[3 marks]

d What properties would be needed in a material used to make a car windscreen?

.... [1 mark]

3 Look at the table and answer the questions below.

Condition / Metal	At start	Dry air	Moist clean air	Moist acidic air
aluminium	shiny silver	shiny silver	shiny silver	dull silver
copper	shiny salmon-pink	shiny salmon-pink	small patches of green on surface	green layer on surface
iron	shiny silver	shiny silver	small patches of brown on surface	lots of brown flakes on surface

a Which metal corrodes the least in moist, clean air?

.... [1 mark]

b Which conditions are likely to cause the most corrosion on a car body?

.... [1 mark]

c Which would be the best material for a car body. Give a your reason for your answer.

.... [2 marks]

1 a Where do the two gases needed to make ammonia in the Haber process come from?

....

.... [2 marks]

G–E

b The Haber process is a reversible reaction. What is the symbol used in an equation to show that it is a reversible reaction.

.... [1 mark]

c Suggest **three** conditions that increase yield in the Haber process.

....

.... [3 marks]

D–C

d Ammonia (NH_3) is made by reacting nitrogen (N_2) and hydrogen (H_2). Construct a balanced symbol equation for this reaction.

.... [2 marks]

2 Look at this graph.

Percentage of ammonia made

(graph: percentage of ammonia made vs pressure in atmospheres, with curves labelled 350 °C, 400 °C, 450 °C)

a How much ammonia is made at a temperature of 400 °C and a pressure of 300 atmospheres?

.... [1 mark]

D–C

b How does pressure affect the yield?

....

.... [1 mark]

G–E

3 a List **three** different factors that affect the cost of making a new chemical.

....

....

.... [3 marks]

G–E

b Explain how one of these factors affects the cost.

.... [1 mark]

c How is the production of ammonia linked to world food production?

....

....

.... [3 marks]

D–C

1 a What is the pH range of an acid?

.... [1 mark]

b What is an alkali?

.... [1 mark]

c What colour does litmus turn with an:

i acid .... [1 mark]

ii alkali .... [1 mark]

2 a What is the name of the process when an acid and alkali react together? Choose from the list below.

    condensation   evaporation   neutralisation   fermentation

.... [1 mark]

b Finish this word equation for neutralisation.

.... + base → salt + ... [1 mark]

c Name the ion that all acids form in solution.

.... [1 mark]

d Explain how the concentration of hydrogen ions is linked to pH.

.... [1 mark]

3 a How does the pH of an acid change when an alkali is added to it?

.... [1 mark]

b **Universal indicator solution** can be used to measure the acidity of a solution. A few drops are added to the test solution and then the colour of the solution is compared to a standard colour chart.

Describe how the colour changes when a strong acid is added to an alkali to neutralise it.

....
....
.... [3 marks]

4 a Write the **word equation** for the reaction between copper carbonate and sulfuric acid.

.... [2 marks]

b Name the **salt** made when:

i sulfuric acid reacts with calcium carbonate   .... [1 mark]

ii nitric acid reacts with potassium hydroxide   .... [1 mark]

iii hydrochloric acid reacts with sodium carbonate .... [1 mark]

iv nitric acid reacts with copper oxide   .... [1 mark]

1 a What do fertilisers do?

.... [1 mark]

b How are fertilisers absorbed?

.... [1 mark]

c Which essential elements do fertilisers provide?

.... [3 marks]

d 'Colourburst' fertiliser contains KNO_3. Which essential element(s) does it contain?

.... [2 marks]

e Why do fertilisers need to be dissolved?

....

.... [1 mark]

2 a What is the benefit of using fertilisers?

.... [1 mark]

b Why can the use of fertilisers sometimes cause problems?

.... [1 mark]

3 a Label the three pieces of apparatus needed in the preparation of a fertiliser by the neutralisation of an acid and an alkali. [3 marks]

b Name two fertilisers manufactured from ammonia.

....

.... [1 mark]

c Fertilisers are made by reacting an acid and an alkali. Name the fertilisers which would be made by reacting:

i nitric acid and ammonia .... [1 mark]

ii phosphoric acid and potassium hydroxide .... [1 mark]

1 a Give two places that salt can be obtained from.

.... [2 marks]

b Name **two** methods of extracting salt from underground deposits found in Cheshire.

....

.... [2 marks]

c Suggest why salt mining can create problems.

....

.... [2 marks]

2 Electrolysis can be used to separate concentrated sodium chloride solution.

a Which two gases are given off in the process?

.... [2 marks]

b Describe a test for one of the gases that does not involve burning.

.... [1 mark]

c In this electrolysis what is formed:

i At the **anode**?

.... [1 mark]

ii At the **cathode**?

.... [1 mark]

iii In the **solution**?

.... [1 mark]

d Why are **inert electrodes** needed?

.... [1 mark]

3 a Give two uses of sodium chloride.

.... [2 marks]

b Give two uses of chlorine.

.... [2 marks]

c Give one use of hydrogen.

.... [1 mark]

d Give one use of the alkali made in the electrolysis of sodium chloride.

.... [1 mark]

e Name the two chemicals which make household bleach.

....

.... [2 marks]

This equipment can be used
to produce a fertiliser.

Label the equipment needed to produce a fertiliser. Predict the names of **three** fertilisers you could produce with nitric acid, sulfuric acid, potassium hydroxide and ammonia.

❗ The quality of written communication will be assessed in your answer to this question.

....
....
....
....
....
....
....
....
....
....
....
....
....
....
....
....
....
....
....
....
....
....
....
....
....
.... [6 marks]

1 a Finish the sentence. Choose words from this list.

degrees Celsius   energy   joules   kilograms   power   watts

Heat is a form of .... and is measured in ....   [2 marks]

b Kelly opens the front door on a very cold morning. Her mother complains that the house is getting cold. Use your ideas about energy flow to explain why the house gets cold.

....

.... [2 marks]

2 a A block of copper of mass 100 g is heated in a Bunsen flame. It takes two minutes to increase the temperature of the copper by 30 °C. How long will it take to increase the temperature of a 200 g block of copper by 60 °C? Choose from this list.

30 seconds   1 minute   2 minutes   4 minutes   8 minutes

.... [1 mark]

b i Finish the sentence.

The energy needed to raise the temperature of 1 kg of a material by 1 °C is known as the

.... [1 mark]

ii Jamie heats 0.5 kg of seawater in a beaker so the temperature increases by 70 °C. The specific heat capacity of seawater is 3900 J/kg °C. How much energy is needed to heat the sea water?

....

....

....

.... [2 marks]

iii The energy supplied is greater than this. Suggest where this additional energy has been used.

....

....

.... [1 mark]

3 a Write down **one** example where energy is transferred but there is no change in temperature.

.... [1 mark]

b What physical quantity is measured in units of J/kg?

.... [1 mark]

1 On a cold winter's day, Jack wears a thick waterproof coat and John wears a thin hoody. Jack feels cold but John stays warm.

Finish the sentences to explain why John stays warm.

The hoody contains .... air. Air is a good .... **[2 marks]**

2 a 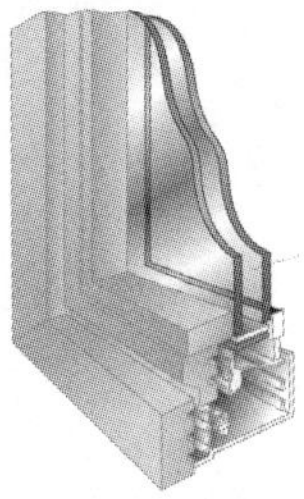 The diagram shows a section through a double-glazed window. Michael says that it is just as effective to use a piece of glass twice the thickness. Use your ideas about energy transfer to explain why double glazing is better.

....

....

.... **[2 marks]**

b New homes are built with insulation blocks in the cavity between the inner and outer walls. The blocks have shiny foil on both sides.

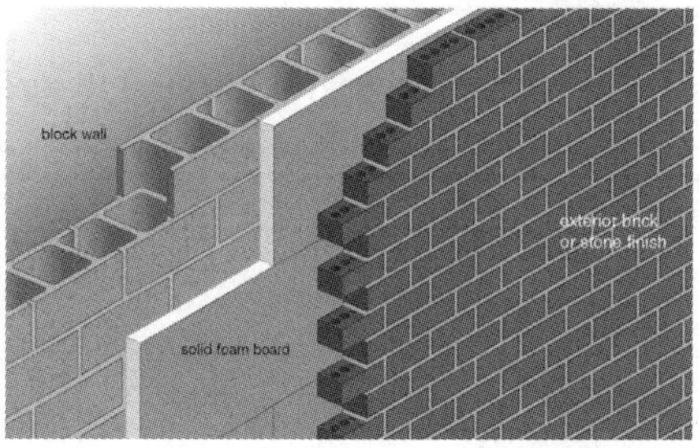

i Explain how the insulation blocks reduce energy transfer by conduction and convection.

block wall

exterior brick or stone finish

solid foam board

....

....

....

.... **[3 marks]**

ii Explain how the shiny foil helps to keep a home warmer in winter and cooler in summer.

....

....

.... **[2 marks]**

c Suggest **one** other way in which energy loss from a home can be reduced.

.... **[1 mark]**

3 Dan heats his house with coal fires. He is told that his fires are 32% efficient.

a Explain what is meant by 32% efficient.

....

.... **[1 mark]**

b Dan pays £9.50 for a 25 kg bag of coal. How much of that money is usefully used in heating his house?

....

....

.... **[2 marks]**

c Suggest why coal fires are so inefficient.

....

.... **[1 mark]**

1 a Light is a transverse wave that travels at a speed of 300 000 km/s.

Write down the names of three other types of transverse wave that travel at 300 000 km/s.

....

....

.... [3 marks]

b The diagram represents a water wave. Water is a transverse wave.

The arrow shows the direction of motion of the wave.

Add another arrow to the diagram to show in which direction the water particles move. [1 mark]

G–E

2 The diagram shows a transverse wave.

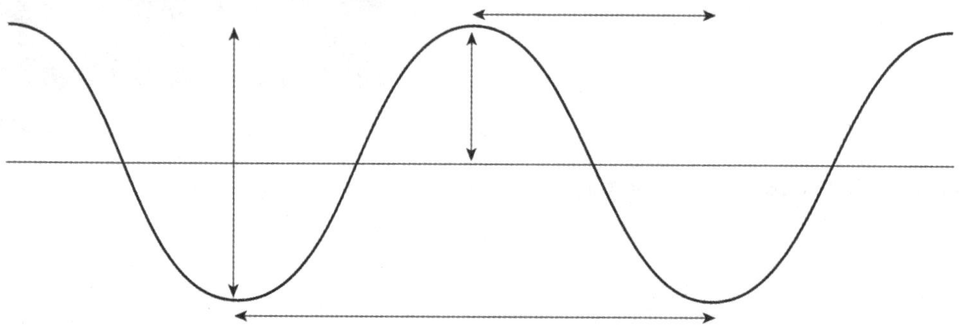

G–E

a Write the letter A next to the arrow which shows the amplitude of the wave. [1 mark]

b Write the letter W next to the arrow which shows the wavelength of the wave. [1 mark]

c What is meant by the frequency of a wave?

....

.... [1 mark]

D–C

3 A simple kaleidoscope has two mirrors inclined at 60°. Draw a ray diagram to show what happens when a ray of light is incident on one mirror with an angle of incidence of 60°.

[2 marks]

D–C

1 Write down one way of sending a message over a long distance without using electricity or radio waves.

.... [1 mark]

2 Why is Morse code an example of a digital signal and not an analogue signal?

....

.... [2 marks]

3 a Finish the sentence to explain what is meant by total internal reflection.

When light is travelling from a .... dense material such as

.... into a .... dense material such as

.... it may be reflected inside the material.

The angle of .... equals the angle of reflection. [5 marks]

b The diagrams show three rays of light travelling from water into air. The three angles of incidence are (x) smaller than the critical angle (y) equal to the critical angle (z) larger than the critical angle.

x   y   z

i Finish the diagrams to show what happens to the rays of light after they meet the water/air boundary. [4 marks]

ii Show clearly, on the correct diagram, the critical angle. Label it c. [1 mark]

4 Lasers are used at supermarket checkouts to read bar codes. Write down two other uses of lasers.

....

.... [2 marks]

1 a Finish the sentences. Choose words from this list.

G-E

absorb   aluminium   induction   infrared   reflect   spectrum   ultraviolet   water

Warm and hot bodies emit .... radiation.

Dark surfaces .... more radiation than light surfaces.

Microwaves are part of the electromagnetic ....

.... molecules absorb microwaves.   **[4 marks]**

b Microwave ovens take less time to cook food than normal ovens. Suggest why.

D-C

....

.... **[1 mark]**

2 a Young people are advised not to use mobiles phones too much. Texting is preferable to using them as a telephone. Why is this advice given to young people?

G-E

....

....

....

.... **[2 marks]**

b Microwaves are suitable to communicate with spacecraft thousands of kilometres away, but mobile phones often cannot receive a signal just a few kilometres from the nearest transmitter. Why do microwave signals seem to work better in space than they do on Earth?

D-C

....

....

....

.... **[2 marks]**

1 a Answer **true** or **false** to each of these statements.

Infrared radiation is part of the electromagnetic spectrum. ....

Passive infrared sensors emit infrared radiation to detect burglars. ....

A remote controller for a CD player emits infrared radiation. .... **[3 marks]**

b The diagram shows a signal displayed on an oscilloscope. What type of signal is it?
Put a ring around the correct answer.

background   digital   on/off   radial   square   **[1 mark]**

c Draw a diagram to represent an analogue signal.

[1 mark]

2 a The change from analogue to digital transmission of television signals began in the United Kingdom in 2009. Write down one advantage of digital television.

....

.... **[1 mark]**

3 A ray of laser light is shone into one end of an optical fibre.

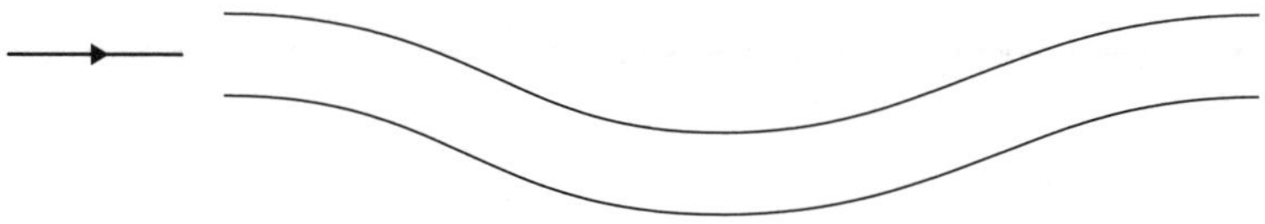

Finish the path of the ray as it passes into, through and out of the optical fibre. **[2 marks]**

1 Radio waves are refracted in the upper atmosphere. What happens to the amount of refraction if the frequency of the radio wave is decreased?

....

.... [1 mark]

2 Jenny is using a mobile phone.

Write down **one** advantage a mobile phone has over a landline house phone.

....

.... [1 mark]

3 When Jenny is watching her television, she notices that there is a faint second picture slightly offset to the main picture.

Finish the sentence to explain why there is this 'ghost' picture.

Choose words from this list.

    absorbed   **dispersed**   **reflected**   **refracted**

The aerial has received a direct signal from the transmitter and a signal that has been

....   [1 mark]

4 Jenny listens to her favourite radio station. Every so often, she notices that she can hear a foreign radio station as well. Put ticks (✓) in the **two** boxes next to the statements that explain why this happens.   [1 mark]

The foreign radio station is broadcasting on the same frequency.   ☐

The foreign radio station is broadcasting with a more powerful transmitter.   ☐

The radio waves travel further because of weather conditions.   ☐

Jenny's radio needs new batteries.   ☐

1 a

rotating drum

The picture shows a device used to detect and measure the strength of earthquakes.

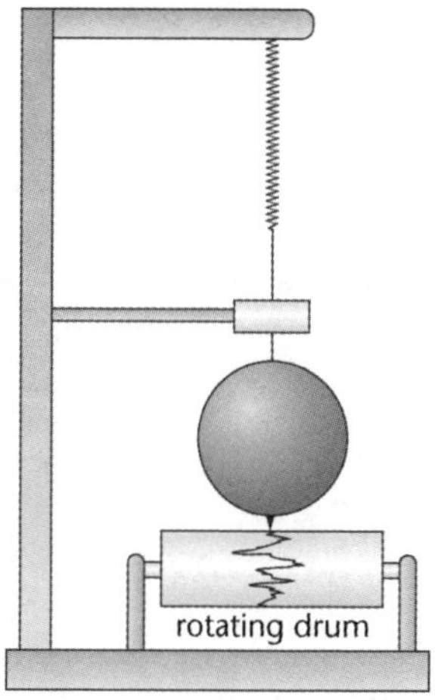

What is the device called? Put a ring around the correct answer.

joulemeter   newtonmeter

seismometer   wattmeter   [1 mark]

b P waves and S waves are two of the waves which travel through the Earth after an earthquake. Finish the table by putting a tick (✓) in the correct box or boxes next to the description of the wave. The first one has been done for you.

description	P wave	S wave
pressure wave	✓	
transverse wave		
longitudinal wave		
travels through solid		
travels through liquid		

[3 marks]

2 Sandy wants to sunbathe and get a good tan.

a What causes Sandy's skin to tan?

....

.... [2 marks]

b Sandy is told that if she goes out in the Sun without sunscreen, she will burn in 15 minutes. How long can she safely sunbathe for if she uses a sunscreen with SPF 20?

....

.... [2 marks]

c The Earth's atmosphere contains a layer of ozone. This layer protects us from the effects of the Sun. When scientists first started measuring the thickness of the ozone layer, their results were unexpected. The layer was thinner than they thought it should be. They replaced all of their instruments. What should scientists do to confirm their results?

....

.... [1 mark]

Many people now use mobile phones.

Jenny is using her mobile to speak to her brother who is working overseas. Describe how electromagnetic radiation is used to transmit a signal from Jenny's phone to her brother's phone. Why do some people think it would be better if Jenny were to text her brother instead?

❶ The quality of written communication will be assessed in your answer to this question.

....
....
....
....
....
....
....
....
....
....
....
....
....
....
....
....
....
....
....
....
....
....
....
....
....
....
....
....
....
....
....
....
.... [6 marks]

1 a Finish the sentence. Choose words from this list.

       chemicals   electricity   heat   light

    A photocell uses .... to produce ....   **[1 mark]**

G–E

b Write down **four** advantages of using photocells.

....

....

....

....   **[4 marks]**

D–C

2 a Solar panels use energy from the Sun to heat water. The solar panel is black. Water from the pipes in the solar panel is stored in a cylinder.

   i Why is the solar panel black?

   ....   **[1 mark]**

   ii How is energy transferred from the water in the pipes to the storage cylinder?

   Put a ring around the correct answer.

       conduction   convection   radiation   **[1 mark]**

b This is a question about passive solar heating.

During the day, energy from the Sun passes through the large window and warms the room.

i How do the walls and floor help in heating the room during the night?

....

....

....   **[1 mark]**

G–E

   ii Why do more houses in Australia have windows facing the North?

   ....

   ....   **[1 mark]**

3 a A wind turbine transfers the kinetic energy of the wind into electricity. What is wind?

....   **[1 mark]**

G–E

b Write down **two** disadvantages and **two** advantages of generating electricity using wind turbines.

Advantages

....

....

Disadvantages

....

....   **[4 marks]**

D–C

1 a The diagram shows a wire connected to an ammeter moving upwards between the poles of a magnet. The needle on the ammeter moves to the right. What happens when the wire is moved downwards between the poles of the magnet?

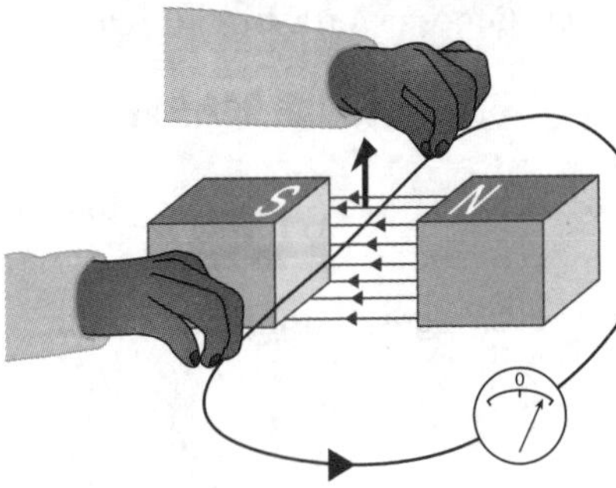

.... **[1 mark]**

b The diagram shows a model dynamo. When the coil is spun, a current is produced.

Write down two ways in which the size of the current can be increased.

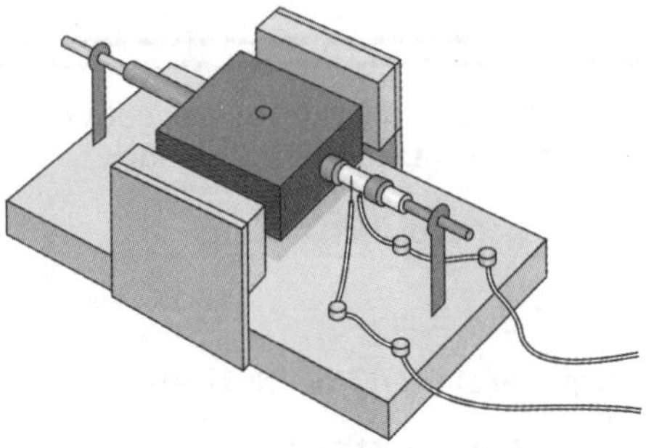

....

....

....

.... **[2 marks]**

c A model generator can be turned by hand. How is the generator at a power station made to rotate?

.... **[1 mark]**

2 The diagram represents how fossil fuels at a power station provide electrical energy for distribution around the country.

| fossil fuel burned | → | water heated to produce steam | → | steam turns turbine | → | turbine turns generator | → | A | → | electricity distributed |

a One step in the process has been missed out. What should be in box A?

.... **[1 mark]**

b How is energy lost from the overhead power lines as electricity is distributed around the country?

....

....

.... **[1 mark]**

c i A power station produces 30 MJ of electrical energy. If the fuel provides 120 MJ of energy, what is its efficiency?

.... **[1 mark]**

ii What are the main causes of inefficient energy transfer in a power station?

.... **[1 mark]**

1 **a** Carbon dioxide is an example of a greenhouse gas. State two other gases which add to the greenhouse effect of the Earth.

....

.... [2 marks]

G–E

b Which of the greenhouse gases is the most significant?

.... [1 mark]

D–C

2 **a** How has deforestation affected the levels of carbon dioxide in the atmosphere?

.... [1 mark]

b Write down **four** natural sources of carbon dioxide.

....

....

....

.... [4 marks]

D–C

c Write down **three** human activities that lead to increased levels of carbon dioxide in the Earth's atmosphere.

....

....

.... [3 marks]

G–E

3 **a** What is global warming?

....

.... [2 marks]

G–E

b Dust in the atmosphere can cause either an increase or a decrease in the Earth's temperature. Explain how this is possible.

Increase in temperature is caused by:

.... [1 mark]

D–C

Decrease in temperature is caused by:

.... [1 mark]

4 **a** How can scientists help to find out what is happening to the Earth's atmosphere?

....

.... [2 marks]

G–E

b Most scientists agree that the average temperature of the Earth is increasing but what do they disagree on?

.... [1 mark]

c On what basis should governments make decisions on what action to take on global warming?

.... [1 mark]

D–C

d Describe **two** consequences of global warming.

.... [2 marks]

1 a Each of the headlamp bulbs in Sammy's car is connected to a 12 V battery. When she switches on the headlamps, a current of 2 A passes through the bulb. Calculate the power rating of the bulb.

....

....

....

.... **[4 marks]**

b In her home, Sammy uses a 2.5 kW kettle for half an hour each day. Electricity costs 12p per kWh. How much does it cost Sammy each day to use her kettle?

....

....

....

.... **[4 marks]**

D–C

2 a What type of fuels are the most common energy sources for power stations?

.... **[1 mark]**

b Some power stations use renewable fuels and others use nuclear fuels.

i Name three examples of renewable fuels that can be used in power stations.

.... **[3 marks]**

ii Name two nuclear fuels.

.... **[2 mark]**

G–E

c List three factors which should be taken into account when deciding on which energy source is best in a particular situation.

....

....

.... **[3 marks]**

D–C

3 The National Grid distributes electricity around the country at 400 000 V.

a Give **two** reasons why such high voltages are used.

.... **[2 marks]**

D–C

b What is the job of a transformer? Put a tick (✓) in the box next to the correct answer.

Change the size of an AC voltage ☐

Change the size of a DC voltage ☐

Change AC into DC ☐

[1 mark]

G–E

1 Answer true or false to each of the following statements about alpha, beta and gamma radiation.

Gamma radiation causes more ionisation than alpha radiation.	
Alpha radiation has a range of a few centimetres in air.	
Beta radiation comes from the nucleus of an atom.	
Beta radiation can be absorbed by a thin sheet of paper.	

[4 marks]

D–C

2 Kelly's teacher is showing the class an experiment using sources of radiation.

a Write down three precautions the teacher should take to protect herself from the effects of the sources.

....

....

.... [3 marks]

G–E

b Describe the effects ionisation can have on the cells of the human body.

....

.... [2 marks]

D–C

3 Gamma radiation is used to sterilise medical instruments. It has other medical uses as well.

a Write down one other medical use for gamma radiation.

.... [1 mark]

b What property of gamma radiation makes it suitable?

.... [1 mark]

D–C

c A source of alpha radiation is used in a smoke alarm. Explain how a smoke alarm works.

....

....

.... [3 marks]

4 Give one advantage and one disadvantage of nuclear power stations.

....

.... [2 marks]

G–E

5 Radioactive waste from power stations must be stored securely for possibly thousands of years.

a Why must it be stored for so long?

.... [1 mark]

D–C

b Describe how low-level radioactive waste is disposed of.

.... [1 mark]

1

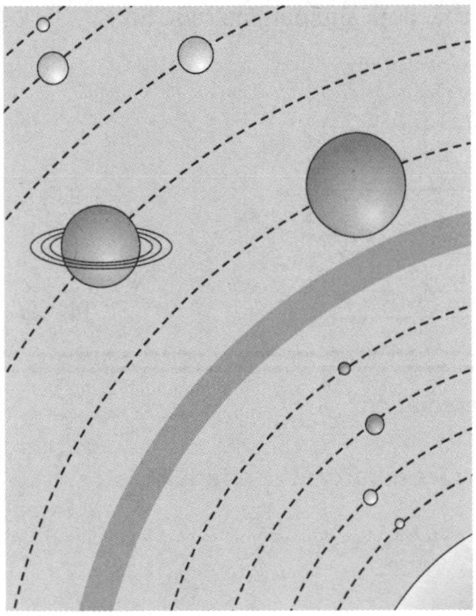

On 24 August 2006, the International Astronomical Union considered a proposal to redefine planets. There would be twelve planets in our Solar System. Ceres, the largest asteroid, would become a planet. Charon, at the moment known as Pluto's moon, would become a 'twin planet' with Pluto. Another planet has recently been discovered beyond Pluto. This proposal was rejected and the decision made that Pluto should no longer be called a planet.

The diagram shows the Sun, the eight planets, Pluto and the asteroid belt.

i Write the letter C on the diagram to show where Ceres orbits.

ii Write the letter P on the diagram to show Pluto.   **[2 marks]**

---

2 a There are many objects in our Solar System. Describe the following objects:

i a star

.... **[1 mark]**

ii a planet

.... **[1 mark]**

iii a meteor

.... **[1 mark]**

b Our Solar System is part of a galaxy and scientists think there may be a black hole at the centre of our galaxy.

i What is a galaxy?

.... **[1 mark]**

ii Describe a black hole.

.... **[1 mark]**

---

3 Scientists are exploring space to find out if there are any other forms of life somewhere in the Universe. Manned and unmanned spacecraft have been sent into space.

a How have scientists tried to find out if there are other life forms without sending a spacecraft into space?

.... **[1 mark]**

b When astronauts work outside a spacecraft, they have to wear special helmets with Sun visors. Why do the helmets need special visors?

.... **[1 mark]**

c Write down **three** things a spacecraft carrying astronauts would have to take on a mission.

....
....
.... **[3 marks]**

1 a An asteroid hit Earth about 65 million years ago. It made a large crater.

What else happened when the asteroid hit Earth? Put ticks (✓) in the boxes next to the **three** correct answers.

All life became extinct. ☐

It got colder. ☐

The first human appeared on Earth. ☐

The Moon was formed. ☐

There were a lot of fires. ☐

Tsunamis flooded large areas. ☐

[3 marks]

b Most asteroids orbit the Sun in a belt between two planets. Which two planets?

....

.... [2 marks]

c Describe **two** pieces of evidence scientists have discovered that support the theory that asteroids have collided with Earth in the past.

....

.... [2 marks]

2 There is evidence to suggest our Moon was a result of the collision between two planets. The iron core of the other planet melted and joined with the Earth's core.

Describe how the Moon was formed as a result of this collision.

....

.... [2 marks]

3 The diagram shows the orbits of two bodies orbiting the Sun. One is a planet, the other is a comet.

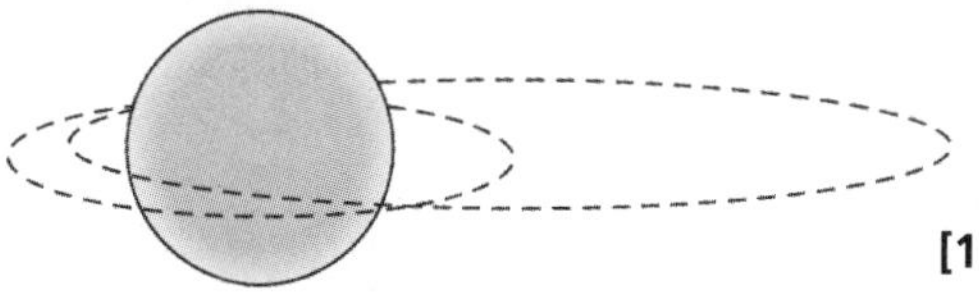

a Label the comet with the letter C. [1 mark]

b When does the tail of a comet become visible?

.... [1 mark]

c How does the speed of a comet change as it orbits the Sun?

.... [1 mark]

4 Scientists are constantly updating information on the paths of near Earth objects (NEO)s.

a What are NEOs?

.... [1 mark]

b Why is it important to constantly monitor the paths of NEOs?

....

.... [2 marks]

1 Scientists think that 15 billion years ago there was a 'Big Bang'.

G–E

a What was the Universe like before the Big Bang?

.... [1 mark]

b What were the first things formed after the Big Bang? Put (rings) around the three correct answers.

copper   helium   hydrogen   iron   protons   uranium   [3 marks]

D–C

c The Universe is expanding. Galaxies in the Universe are moving at different speeds. Which galaxies are moving the fastest?

.... [1 mark]

2 Models of our own Solar System have changed over time.

G–E

a Who suggested the idea that the planets orbit the Sun?

.... [1 mark]

b How did Galileo help to support this model of our Solar System?

.... [1 mark]

3 New stars are being formed all the time.

G–E

a How do stars start to be formed?

.... [1 mark]

The end of a star's life depends on how big it is.

b What happens to a medium-sized star, like our Sun, at the end of its life?

D–C

....
....
....
.... [4 marks]

P2 Living for the future (energy resources)

With the increasing demand for energy in 2008, Britain was deciding whether or not to build new nuclear power stations. Many people are concerned about the dangers of nuclear power.

Describe how nuclear power stations produce electricity from nuclear fuels. Nuclear power stations are costly to build and maintain but what are the other reasons that make some people concerned about the use of these types of power stations?

❶ The quality of written communication will be assessed in your answer to this question.

....
....
....
....
....
....
....
....
....
....
....
....
....
....
....
....
....
....
....
....
....
....
....
....
....
....
....
....
....
....
....
.... [6 marks]

1 a Draw straight lines to show the position in the cell and the job of chromosomes and mitochondria. You should draw **four** lines only.

Position in the cell		Job in the cell
In the cytoplasm	Chromosomes	The site of respiration
On the cell membrane	Mitochondria	Contain the genetic code
In the nucleus		Control what enters the cell

[2 marks]

b Muscle cells contain many mitochondria. Explain why muscle cells need so many mitochondria.

....

.... [2 marks]

2 a Finish the sentences about chromosomes. Choose the best words from the list:

    base   cytoplasm   fat   gene   nucleus   protein

A section of a chromosome is called a ....

Each section is a code for making a .... [2 marks]

b The diagram shows part of a DNA molecule.

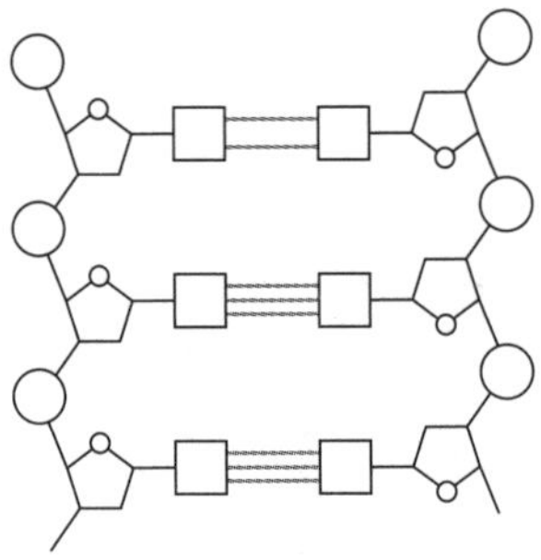

    i On the diagram, put a circle around **one** base. [1 mark]

    ii What term is usually used to describe the shape of a DNA molecule?

    .... [1 mark]

    iii Cells make copies of sections of DNA. These copies then pass out of the nucleus into the cytoplasm. Explain why this happens.

    ....

    .... [2 marks]

3 Write down the names of the two scientists who discovered the structure of DNA.

.... [1 mark]

b A scientist called Erwin Chargaff worked out the percentage of the four bases A, G, C and T in different organisms. The table shows some of his results.

	Percentage of each DNA base			
	A	G	C	T
humans	29	21	21	29
bacteria	24	26	26	24

Describe the patterns shown by the scientist's results.

.... [2 marks]

B3 Living and growing

1 **a** Draw straight lines to join each protein to its correct function.

Protein	Function
collagen	a carrier protein
haemoglobin	a hormone
insulin	a structural protein

[2 marks]

G–E

b Write down the name of the subunits that make up protein molecules.

.... [1 mark]

2 **a** What is meant by the term 'enzyme'?

....

.... [2 marks]

G–E

b What is meant by the term 'active site'?

....

.... [2 marks]

c The enzyme amylase breaks down starch. Explain why amylase would not break down proteins.

.... [2 marks]

d Look at the graph. It shows the effect of temperature on the enzyme amylase.

rate of reaction (arbitrary units) vs temperature (°C)

i Describe the pattern shown in the graph.

....

....

.... [2 marks]

D–C

ii Write down the optimum temperature for this enzyme.

....

.... [1 mark]

3 Genetic conditions such as cystic fibrosis are caused by changes in genes.

a What name is given to a change in a gene?

.... [1 mark]

G–E

b Changes to genes can happen spontaneously or can be caused by factors in the environment. Write down one environmental factor that increases the chance of a change in a gene.

.... [1 mark]

c Why are changes to genes important for evolution?

....

.... [2 marks]

D–C

1 a A horse is waiting to run in a race. Write down **two** processes occurring in the horse's body that need energy from respiration.

....

.... [2 marks]

b Glycogen is a carbohydrate that is stored in the muscles of the horse. It can be broken down to release glucose, which is used in aerobic respiration.

Write down one other substance that is needed for aerobic respiration.

.... [1 mark]

c The horse starts to run.

Complete the balanced symbol equation for aerobic respiration in the horse.

$C_6H_{12}O_6$ + ... → .... + ... [2 marks]

d As the horse runs, it starts to breathe faster. Why is this?

....

.... [2 marks]

e The graph shows the lactic acid concentration in the horse's blood as it runs at different speeds.

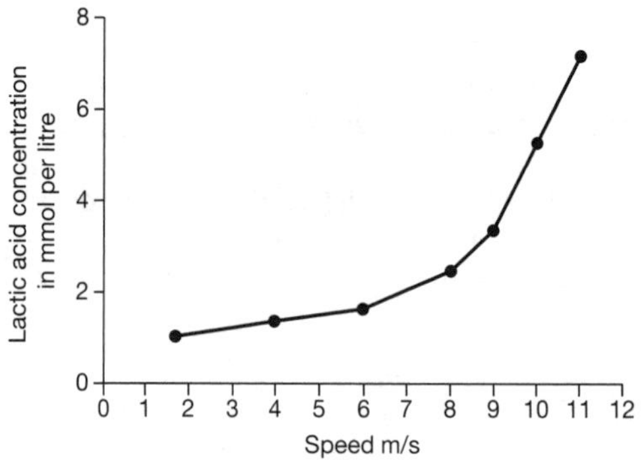

i Describe the shape of the graph.

....

.... [2 marks]

ii Explain the pattern shown by the graph.

....

....

.... [3 marks]

iii Horses are more likely to damage their muscles when their blood lactic acid concentration is above 4 mmol per litre.

How fast can this horse run before this level is reached?

.... [1 mark]

2 Suggest **one** way that the rate of respiration of an animal can be measured.

....

.... [1 mark]

B3 Living and growing

1 Amoebae are unicellular organisms.

a What does this mean?

.... [1 mark]

G–E

b Humans are multicellular organisms. There can be advantages to being multicellular rather than unicellular. Explain why.

....

.... [2 marks]

D–C

2 a Write down the name of the type of cell division that makes new **body** cells.

.... [1 mark]

G–E

b Look at the statements about cell division. Put a tick (✓) next to each statement that refers to the type of cell division that makes new body cells.

The new cells are diploid. ☐

Four new cells are made. ☐

The new cells contain 23 chromosomes. ☐

The new cells show variation. ☐

Before cells divide, DNA replication takes place. ☐ [2 marks]

D–C

3 Tom and Jen want to have a baby. Fill in the gaps in these sentences. Choose words from this list:

double   **fertilisation**   **half**   **meiosis**   **mitosis**   **pregnancy**

Each of Tom's sperm is produced by a process called ....

This makes sure that the number of chromosomes is .... the number in a body cell.

To produce a baby, Tom's sperm must join with one of Jen's eggs.

This process is called .... [3 marks]

G–E

4 Scientists have discovered a mutation in the DNA of mice. They have found that this change makes the mice produce sperm without an acrosome.

The sperm that are produced without an acrosome **cannot** fertilise an egg. Explain why.

....

....

.... [2 marks]

D–C

1 a Draw a straight line from each part of the blood to the job it does.

Part of the blood	The job it does
white blood cell	transports oxygen
red blood cell	helps clot blood
platelet	defends against disease

[2 marks]

b Red blood cells are adapted to do their job. They are disc shaped and have no nucleus.
Explain how these adaptations help them do their job.

Disc shaped: .... [1 mark]

No nucleus: .... [1 mark]

c Write down the name of the chemical that makes red blood cells red.

.... [1 mark]

2 Different types of blood vessels transport blood around the body.

a Write down the name of the three main types of blood vessel.

.... [3 marks]

b Describe the role of these blood vessels in circulating blood around the body.

....
....
.... [3 marks]

3 Look at the diagram of the heart.

a Where does the right side of the heart pump blood to?

.... [1 mark]

b On the diagram of the heart, label the bicuspid valve and the aorta. [2 marks]

c The left ventricle has a thicker wall than the right ventricle. Explain why.

....
.... [2 marks]

B3 Living and growing

1 a Circle two parts of a plant cell that provide support.

cell membrane　　　**cell wall**　　　**cytoplasm**　　　**nucleus**　　　**vacuole**　　　[2 marks]

b Look at the diagram of a bacterial cell.

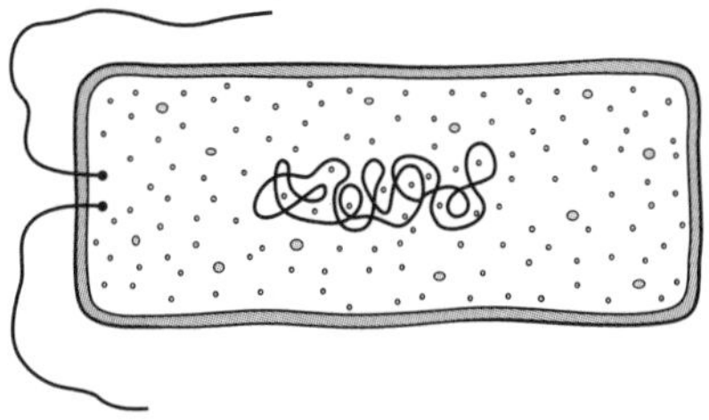

i How does the size of this cell compare to the size of a human cheek cell?

.... [1 mark]

ii Apart from size, write down two differences between the structure of a bacterial cell and a human cheek cell.

.... [2 marks]

G–E

D–C

2

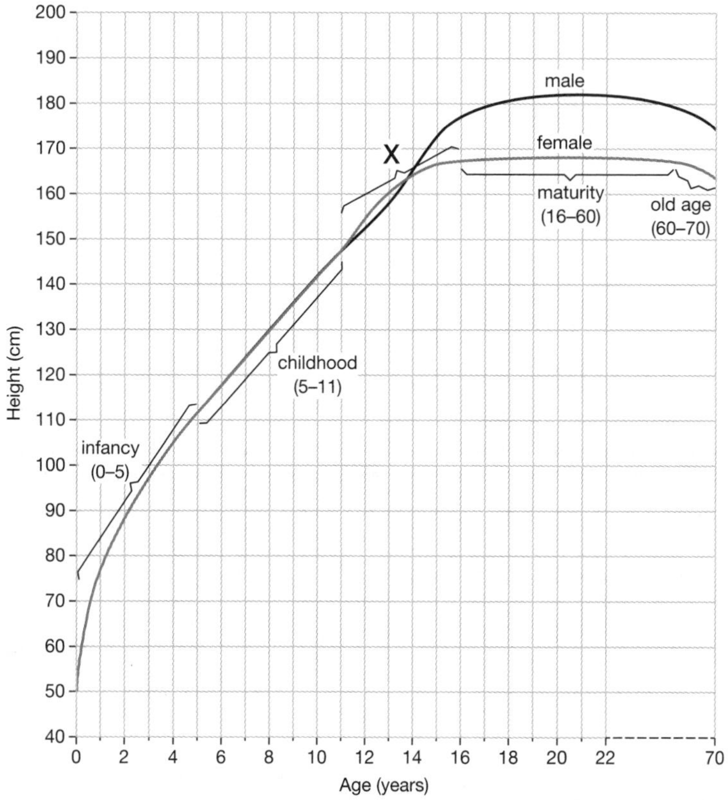

The graph shows the growth curve for a boy and a girl.

a During which phase of growth does the girl grow the fastest?

.... [1 mark]

b Which growth phase is marked as **X** on the graph?

.... [1 mark]

c Suggest the reason for the difference in the height between a girl and a boy at age 13.

....

....

....

....

....

.... [2 marks]

G–E

D–C

3 For a fertilised egg to grow into an embryo the cells need to divide and change.

a What name is given to the way simple cells change into specialised cells such as blood cells? Put a ring around the correct answer.

cell development　　　**cell differentiation**　　　**cell growth**　　　**cell mutation**　　　[1 mark]

b Write down the name that is given to cells in the embryo that can form specialised cells.

.... [1 mark]

G–E

D–C

4 Animals grow in the early stages of their lives. All parts of an animal are involved in growth.

a Rewrite the two sentences to show how plants grow.

.... [2 marks]

b What is a meristem? .... [1 mark]

G–E

D–C

1 Bill grows apples. He uses selective breeding to produce large apples that are resistant to disease and green in colour.

The table below shows some information about different varieties of apples.

	Variety		
	A	B	C
Colour	green	red	green
Resistant to disease	no	yes	no
Large fruit	yes	no	no
Sweet fruit	yes	no	no

a Which two varieties should Bill use in his breeding program?

....

.... **[2 marks]**

b Describe the process of selective breeding.

....

....

.... **[3 marks]**

c Selective breeding in animals can lead to inbreeding. What is inbreeding?

....

.... **[1 mark]**

2 Beta-carotene is found in carrots but not rice.

a The gene for beta-carotene can be transferred from carrots to rice. What is the name of this process?

.... **[1 mark]**

b Suggest why this transfer might be useful.

....

.... **[2 marks]**

c Suggest one possible risk of genetic engineering.

.... **[1 mark]**

3 Scientists hope to be able to transfer genes into people with genetic disorders.

a What might this transfer be used for?

.... **[1 mark]**

b What name is given to this type of process?

.... **[1 mark]**

1 Dolly the sheep was a clone.

a What is a clone?

.... [1 mark]

b Some humans are **natural clones**.

What name is given to two people who are natural clones of each other?

.... [1 mark]

G–E

c Cows can be cloned using **nuclear transfer**.

What is meant by nuclear transfer?

....

.... [2 marks]

d Scientists are hoping to solve organ transplant problems by cloning pigs.

i Suggest how the cloning of pigs could help solve organ transplant problems.

....

.... [2 marks]

D–C

ii Suggest one other reason that people might want to clone pigs.

.... [1 mark]

2 Strawberry plants can reproduce sexually or asexually.

plantlet

a The table shows some of the stages of asexual reproduction in a strawberry plant.

Buds grow on the runners	A
The parent plant grows long stems called runners	B
The runners die	C
The plantlets on the runners grow roots	D
The buds on the runners grow leaves and become plantlets	E

Put the stages in the correct order. The first one has been done for you.

B				

G–E

b Write down **one** advantage and **one** disadvantage to a gardener of reproducing strawberry plants asexually.

....

.... [2 marks]

D–C

Collagen is an important protein in the body.

Sometimes, disorders involving collagen cause problems in the skin and joints. They can also cause blood vessels to weaken and arteries may burst. These disorders can be inherited.

Explain how an inherited disorder can affect collagen and suggest why it causes these particular problems.

❗ The quality of written communication will be assessed in your answer to this question.

....

....

....

....

....

....

....

....

....

....

....

....

....

....

....

....

....

....

....

....

....

....

....

....

....

....

....

....

[6 marks]

....

1 a Describe how the distribution of organisms within a habitat may be affected by the presence of other living organisms.

....

.... [2 marks]

b Look at the kite diagram showing the distribution of different species of barnacles on rocks on a seashore.

i Which species is most abundant?

.... [1 mark]

ii Which species were found at a shore height of 1 m?

.... [1 mark]

2 Which of these ecosystems is natural? Put a tick (✓) in the box next to the correct example.

| artificial lake | ☐ | fish farm | ☐ |
| forestry plantation | ☐ | native woodland | ☐ | [1 mark]

3 Charlie uses a pooter to catch ladybirds in his garden.

a Explain what a pooter is and how it is used.

....

.... [3 marks]

b Charlie catches 50 ladybirds and marks them with a white dot and releases them. The next day he catches 60 ladybirds, but only 10 have a white dot.

i What is the term used to describe this method of estimating population size?

.... [1 mark]

ii Calculate the estimated population size using the formula:

$$\text{population size} = \frac{\text{number in 1st sample} \times \text{number in 2nd sample}}{\text{number in second sample previously caught}}$$

.... [2 marks]

iii Why is this only an estimate?

.... [1 mark]

1 a i Complete the word equation for photosynthesis.

[   ] + [   ] → [glucose] + [   ]   [3 marks]

ii What form of energy is required for photosynthesis?

.... [1 mark]

b i Complete the table to show what happens to glucose in a plant.

Changed to:	Used for:
cellulose	
	storage
proteins	
fats and oils	

[4 marks]

ii If glucose is not changed to other substances, what is it used for?

.... [1 mark]

2 What were van Helmont and Priestley's important contributions in understanding photosynthesis?

van Helmont ....

Priestley .... [2 marks]

3 a Plants grow faster in the summer than in winter. Explain why.

....

.... [3 marks]

b Look at the diagram of a greenhouse.

shades removed from ceiling to allow maximum light — ventilation — paraffin heater producing carbon dioxide — watering system

Describe how the conditions necessary for photosynthesis are achieved.

....

....

.... [3 marks]

1 Plant leaves contain many different types of cells. Explain why some of these cells do not contain chloroplasts.

....

.... [2 marks]

2 Materials involved in photosynthesis enter and exit the plant. Complete the sentences to explain where these materials enter and exit.

Water enters plants through their ....

Carbon dioxide enters plants through .... in the leaves.

Oxygen exits plants through .... in the leaves. [3 marks]

3 a Sam grows tomato plants. Some plants have larger leaves than others. Suggest why having large leaves is important to plants.

....

.... [3 marks]

b Complete the table showing leaf adaptations and their uses.

Adaptation:	Uses:
thin leaves	
	to absorb light from different parts of the spectrum
	support and transport

[3 marks]

4 Look at the diagram of the inside of a leaf.

a Name the parts labelled **A**, **B**, and **C**.

A ....

B ....

C ....

[3 marks]

b An insect called a leaf miner burrows through the inside of leaves and eats the cells.

i Name two types of cells it will eat.

.... [2 marks]

ii Explain why the insect will cause colourless lines on the leaves.

.... [2 marks]

1 Ruby has geranium plants on her windowsill. The leaves begin to wilt.

a Describe what the leaves look like.

.... [1 mark]

b Suggest why the leaves have wilted.

.... [1 mark]

c Which part of a plant cell gives it most support? Draw a ring around the correct answer.

cell wall   **cell membrane**   **cell vacuole**   [1 mark]

---

2 In photosynthesis, gases enter and exit plant leaves. Complete the sentences to explain this.

Use words from this list:

carbon dioxide diffusion enter exit less more osmosis oxygen outside

The movement of gases is due to the process of ....

Photosynthesis uses up .... inside the leaves so there will be

.... outside than inside. Therefore this gas will

.... the leaves.   [4 marks]

---

3 Ruby opens a bottle of perfume. The perfume evaporates. After a few seconds, Ruby smells the perfume.

She draws a diagram to explain this.

molecules of air

perfume molecules

perfume

Position of molecules on opening bottle

Position of molecules after a few seconds

a Using the same symbols, draw in the second box the position of molecules where you expect them to be after a few seconds.   [2 marks]

b Name the process involved in these changes. .... [1 mark]

c Explain how differences in concentration cause these changes.

....

.... [2 marks]

---

4 Complete the definition of osmosis.

Osmosis is the movement of .... across a

.... membrane. It takes place from an area of a

.... solution to an area of a

.... solution. The movement is a consequence of the

.... movement of individual particles.   [5 marks]

1 a Complete the sentences about water in plants.

A plant absorbs water through its ....

The water travels up the stem to the ....

The evaporation of water from plants is called ....

Healthy plants must keep their water loss and water uptake .... **[4 marks]**

b The rate of evaporation of water from plants will vary, depending on conditions such as the amount of light. Name **two** other conditions that will change the rate of evaporation.

1 ....

2 .... **[2 marks]**

G–E

2 Look at the diagram of the lower leaf surface of two leaves kept in different conditions.

stoma closed   stoma open

leaf A   leaf B

a Count the number of open stomata in Leaf A and Leaf B.

Open stomata in Leaf A: ....

Open stomata in Leaf B: .... **[1 mark]**

b Which leaf has been kept in dark conditions? Explain your answer.

....

.... **[2 marks]**

D–C

3 Some leaves from the same plant were smeared with petroleum jelly in different ways, weighed and suspended from a string. The experiment was left for two hours and the leaves were reweighed.

Leaf	Treatment	Initial weight g	Final weight g	Weight change g
1	No petroleum jelly used	10.9	7.2	3.7
2	Petroleum jelly on both leaf surfaces	9.8	9.2	0.6
3	Petroleum jelly on lower surface	10.1	9.1	1.0
4	Petroleum jelly on upper surface	9.9	7.2	2.7

a What was the main cause of weight loss?

.... **[1 mark]**

b Use the data to show that most stomata are in the lower surface.

....

.... **[2 marks]**

D–C

1 Ben buys a sack of fertiliser. It has information on its N, P and K content.

a What do the letters N, P and K stand for?

N: ....

P: ....

K: .... [3 marks]

b Name **one** other mineral that is important for plants.

.... [1 mark]

c i The 20 kg sack of fertiliser shows the NPK content as 3 : 6 : 9.

Work out the mass of each component. (Assume the sack only contains fertiliser.)

N: ....

P: ....

K: .... [3 marks]

ii Ben grows strawberries on 5 m^2 of his allotment. The strawberries need N fertiliser at 0.05 kg/m^2.

Will his sack of fertiliser provide this? Explain your answer.

....

.... [2 marks]

G–E

2 Look at the diagram of plants grown in conditions lacking certain minerals.

poor growth, yellow leaves

poor root growth, discoloured leaves

poor fruits and flowers, discoloured leaves

yellow leaves

A   B   C   D

a Write down the mineral that each plant is lacking.

Plant A: ....   Plant B: ....

Plant C: ....   Plant D: ....

[4 marks]

b Plants lacking nitrogen are usually smaller than usual. Explain why.

.... [2 marks]

....

D–C

1 Decay is important for plant growth. Which statement gives the correct explanation for this?

Put a tick (✓) in the box next to the correct answer.

Plants need water to grow ☐   The chemicals in dead organisms are recycled ☐

Decay uses up oxygen ☐   Plants use decay in transpiration ☐   **[1 mark]**

G–E

2 Ali wants to make garden compost. He wants his garden waste to decay quickly.

a What is compost?

.... **[1 mark]**

G–E

b Ali wants his garden waste to decay quickly. What conditions should he provide?

Put a tick (✓) in the box next to each correct answer.

A temperature of 25 °C ☐   Plenty of carbon dioxide ☐

A temperature of 75 °C ☐   Plenty of oxygen ☐

A large amount of water ☐   **[2 marks]**

c Name **three** useful animals he would expect to find in his compost heap.

1 ....

2 ....

3 .... **[3 marks]**

D–C

d Suggest different views people may have on having compost heaps in their garden.

....

.... **[2 marks]**

e Explain why detritivores are important in decay.

....

.... **[2 marks]**

3 Ali uses a refrigerator and freezer for his garden produce.

a Write down two other ways of preserving food.

....

.... **[2 marks]**

G–E

b Explain how putting foods in:

i a refrigerator at 5 °C

ii a deep freeze at –22 °C

will help to keep them for longer.

....

....

....

.... **[4 marks]**

D–C

1 Draw straight lines to link up each pesticide to its correct function.

Pesticide	Function
insecticide	kills weeds
	helps to grow fungi
fungicide	kills insects
herbicide	kills fungi

[3 marks]

2 Look at the information about percentage losses of various crops.

Crop	Percentage loss due to weeds	Percentage loss due to disease	Percentage loss due to insects
rice	10	9	27
maize	13	9	13
wheat	9	9	5
sugar cane	15	19	19

a Which crop suffers the highest total percentage loss?

.... [1 mark]

b Which crop would benefit most from the use of pesticides? Explain your answer.

....

.... [2 marks]

3 In 1935 large cane toads from South America were introduced into Queensland in Australia to control insect pests that attack sugar cane plants.

The toads had little effect on the insect pest population. They have now spread into nearly all areas of Australia and have no natural predators. They are a pest, eating the smaller native toad and causing a great deal of damage. Because of their large size and poisonous skin, they have few predators. The insect pests are now controlled by insecticides.

a This is an example of introducing an organism into an ecosystem to control another organism. What is this method called?

.... [1 mark]

b Suggest why the introduction of cane toads was thought to be better than using insecticides.

....

.... [2 marks]

c Suggest why the introduction of cane toads was not a success.

....

.... [2 marks]

4 The use of crop rotation is important in farming.

a Explain what crop rotation is and how it works.

....

.... [3 marks]

b Clover is a nitrogen-fixing plant. At the end of the year it is ploughed back into the ground. Explain why.

....

....

.... [3 marks]

Greek philosophers suggested that food for plants was produced in the soil and was absorbed by the roots.

A Dutch chemist called van Helmont carried out an experiment to investigate this.

He planted a young willow tree in a tub of soil, which was watered for five years. The soil was covered to prevent dust getting in.

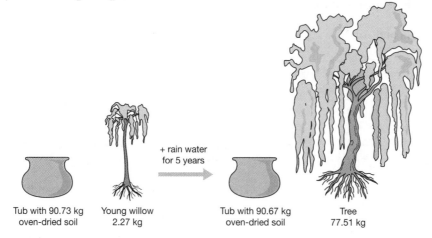

Tub with 90.73 kg   Young willow   + rain water for 5 years   Tub with 90.67 kg   Tree
oven-dried soil   2.27 kg   oven-dried soil   77.51 kg

Use these results to show how this experiment changed our understanding of how plants produce food.

❷ The quality of written communication will be assessed in your answer to this question.

....

....

....

....

....

....

....

....

....

....

....

....

....

....

....

....

....

....

....

.... [6 marks]

1 Give an example of a slow reaction.

.... [1 mark]

2 a Jo and Sam measure the rate of reaction between magnesium and acid. They collect the gas given off. What piece of apparatus will they use to:

i hold the magnesium and acid?

.... [1 mark]

ii collect the gas?

.... [1 mark]

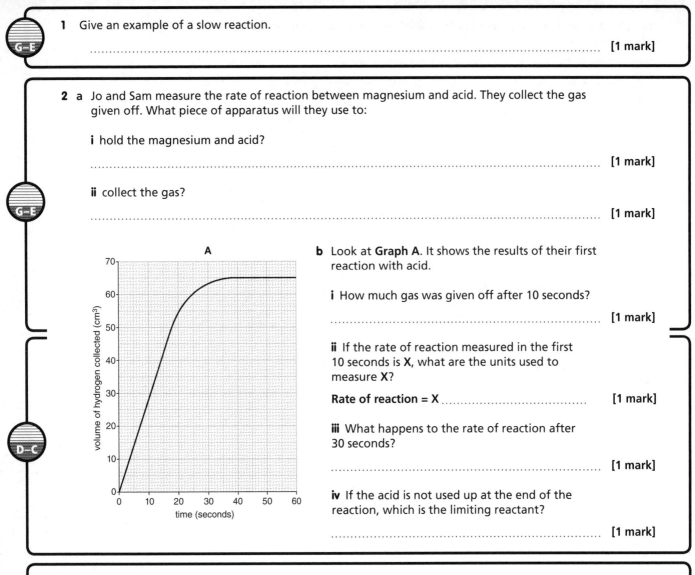

b Look at **Graph A**. It shows the results of their first reaction with acid.

i How much gas was given off after 10 seconds?

.... [1 mark]

ii If the rate of reaction measured in the first 10 seconds is **X**, what are the units used to measure **X**?

Rate of reaction = X .... [1 mark]

iii What happens to the rate of reaction after 30 seconds?

.... [1 mark]

iv If the acid is not used up at the end of the reaction, which is the limiting reactant?

.... [1 mark]

3 Look at **Graph B**.

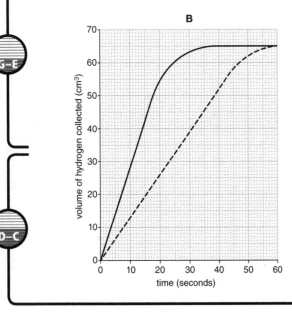

a How long did it take to produce 20 cm³ of gas in reaction 2?

.... [1 mark]

b Look at the gradients of the two graphs. Which reaction was faster?

.... [1 mark]

c At what time does the second line show a change to a slower rate of reaction?

.... [1 mark]

C3 Chemical economics

Rate of reaction (2)

1 Look at **Graph A.** Joe measures the rate of reaction between magnesium and acid at 30 °C and plots it on the graph.

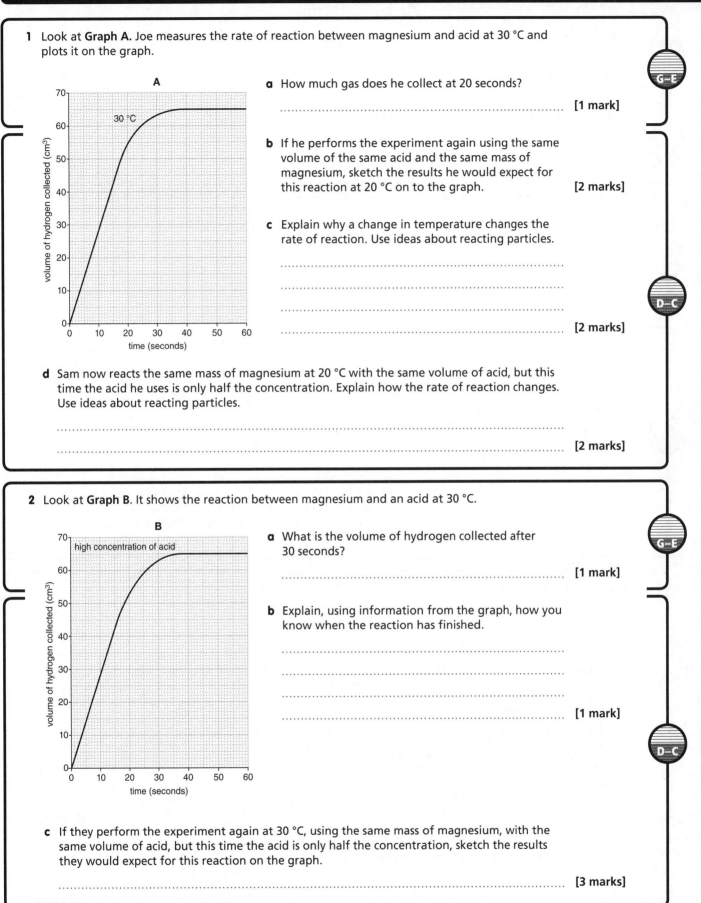

G–E

a How much gas does he collect at 20 seconds?

.... **[1 mark]**

b If he performs the experiment again using the same volume of the same acid and the same mass of magnesium, sketch the results he would expect for this reaction at 20 °C on to the graph. **[2 marks]**

c Explain why a change in temperature changes the rate of reaction. Use ideas about reacting particles.

....

....

....

.... **[2 marks]**

D–C

d Sam now reacts the same mass of magnesium at 20 °C with the same volume of acid, but this time the acid he uses is only half the concentration. Explain how the rate of reaction changes. Use ideas about reacting particles.

....

.... **[2 marks]**

2 Look at **Graph B.** It shows the reaction between magnesium and an acid at 30 °C.

G–E

a What is the volume of hydrogen collected after 30 seconds?

.... **[1 mark]**

b Explain, using information from the graph, how you know when the reaction has finished.

....

....

....

.... **[1 mark]**

D–C

c If they perform the experiment again at 30 °C, using the same mass of magnesium, with the same volume of acid, but this time the acid is only half the concentration, sketch the results they would expect for this reaction on the graph.

.... **[3 marks]**

Rate of reaction (3)

1 a What happens during an explosion?

....

.... [1 mark]

b Why are fine powders, such as custard powder, dangerous in large quantities in factories?

....

.... [3 marks]

2 In the reaction of calcium carbonate and hydrochloric acid, carbon dioxide is given off. Look at graph below. It shows the reaction between calcium carbonate, in small lumps, and hydrochloric acid.

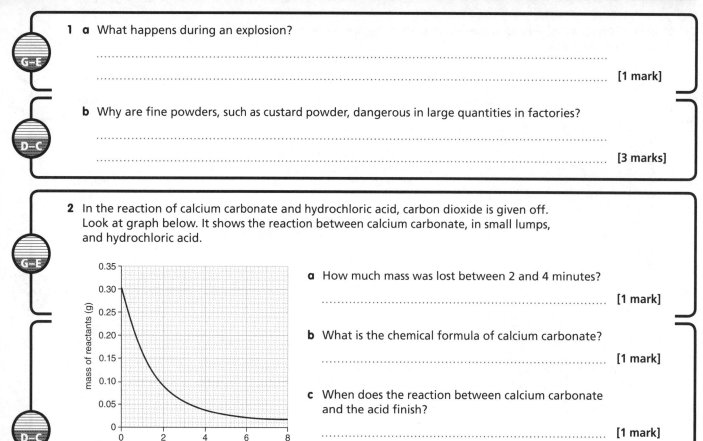

a How much mass was lost between 2 and 4 minutes?

.... [1 mark]

b What is the chemical formula of calcium carbonate?

.... [1 mark]

c When does the reaction between calcium carbonate and the acid finish?

.... [1 mark]

d Sketch on the graph the results you would expect to see for this reaction at the same temperature, using the same volume of the same acid and the same mass of calcium carbonate, but this time in powder form. [2 marks]

3 a Describe what a catalyst is and how much is needed in a reaction.

....

.... [3 marks]

b Jo and Akira have collected results for the reaction between 1.5 g zinc and sulfuric acid. In their next experiment they add 0.1 g of copper powder.

Time in seconds	10	20	30	40	50
volume of gas with no substance cm^3	18	30	60	100	100
volume of gas with copper powder cm^3	70	100	100	100	100

i At what time did each reaction stop?

.... [1 mark]

ii Using the data in the table, compare the rates of the two reactions.

....

.... [2 marks]

1 a How many different elements are there in KNO_3?

.... [1 mark]

b How many atoms are there in KNO_3?

.... [1 mark]

G–E

c Work out the relative formula mass of KNO_3.

.... [1 mark]

A_r (K = 39, O = 16, N = 14)

d How many atoms are there in the formula of calcium hydroxide, $Ca(OH)_2$?

.... [1 mark]

D–C

e Work out the relative formula mass of $Ca(OH)_2$.

A_r (Ca = 40, O = 16, H = 1)

.... [1 mark]

2 a What is the principle of the conservation of mass?

.... [1 mark]

G–E

b Show that when sodium hydroxide reacts with nitric acid to make sodium nitrate and water that mass is conserved.

$NaOH \ + \ HNO_3 \ \rightarrow \ NaNO_3 \ + \ H_2O$

A_r (Na = 23, O = 16, H = 1, N = 14)

D–C

....

.... [2 marks]

3 a When calcium carbonate is heated, it turns into calcium oxide and carbon dioxide. If 50 g of calcium carbonate is heated, it produces 28 g of calcium oxide. How much carbon dioxide is made?

.... [1 mark]

b 1.2 g of magnesium heated in oxygen makes 2.0 g of magnesium oxide.

i Work out how much magnesium oxide is made from 24.0 g of magnesium.

.... [1 mark]

G–E

ii How much magnesium is needed to make 8.0 g of magnesium oxide.

.... [1 mark]

c Zinc reacts with hydrochloric acid to make zinc chloride.

$Zn \ + \ 2HCl \ \rightarrow \ ZnCl_2 \ + \ H_2$

Jo needs to make a certain amount of zinc chloride. How will she make sure that the amount of zinc she uses is the limiting factor?

.... [3 marks]

D–C

d Zinc carbonate ($ZnCO_3$) decomposes on heating to give zinc oxide (ZnO) and carbon dioxide. Write a balanced symbol equation for this reaction.

.... [2 marks]

1 a Leo and Lesley want to make some crystals. They predict that they will make 42 g. They may not make as much as they predict. List three reasons why.

....

....

.... [3 marks]

b Leo and Lesley have now made some crystals of magnesium sulfate, although they have not made as much as they had hoped to. They wanted to make 42 g but only made 28 g.

i What was their **actual yield**? .... [1 mark]

ii What was their **predicted yield**? .... [1 mark]

c i How could they calculate their **percentage yield**?

....

.... [1 mark]

ii Complete their calculation.

....

.... [1 mark]

2 a Leo and Lesley now make some sodium nitrate. The atom economy will not be 100%.

i What does atom economy measure?

.... [1 mark]

ii What does 'The atom economy will not be 100%' mean?

.... [1 mark]

iii Why is it important to have a high atom economy?

.... [1 mark]

b Look at the table below and answer these questions.

Reaction	A	B	C	D
percentage yield	64	72	33	61
atom economy	72	32	64	56

i Which reaction gives the highest percentage yield?

.... [1 mark]

ii Which reaction is the most green and produces the smallest amount of waste product?

.... [1 mark]

c Sodium nitrate is made using sodium hydroxide and nitric acid (water is an undesired product).

$$NaOH + HNO_3 \rightarrow NaNO_3 + H_2O$$

A_r (Na = 23, O = 16, H = 1, N = 14)

i Calculate the relative formula mass of both products.

.... [2 marks]

ii Using the formula for atom economy, calculate the atom economy for sodium nitrate.

....

.... [3 marks]

1 a What is an exothermic process?

.... [1 mark]

b What is an endothermic process?

.... [1 mark]

c Todd and Terri add two solutions together in a beaker. How can they measure whether the reaction between the two solutions is endothermic or exothermic?

.... [1 mark]

d Choose the best correct word from the list to complete sentence (i).

Choose another correct word to complete sentence (ii).

    catalytic   **exothermic**   **endothermic**   **relative**

  i Bond breaking is a process that is .... [1 mark]

  ii Bond making is a process that is .... [1 mark]

2 Todd and Terri want to see which of four fuels releases the most energy. Their friends have did some experiments and got these results.

Fuel	Temperature at start in ˚C	Temperature at end in ˚C	Temperature change in ˚C
A	20	35	
B	19	45	
C	20	37	
D	18	42	

a Which fuel gave out most energy? ....

How do you know?

.... [1 mark]

b Todd and Terri want to compare the energy transferred by two fuels, **A** and **B**. Describe how they can do this by using the apparatus in the diagram.

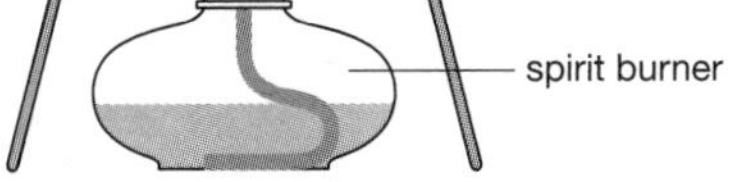

100 g water

spirit burner

....

....

....

....

....

....

....

.... [5 marks]

c Fuel **A** heats the water (100 g) from 20 ˚C to 50 ˚C. They use the formula:

Energy transferred (J) = mass of water × specific heat capacity × temperature change

Work out how much energy Fuel **A** has transferred (specific heat capacity of water is 4.2 J/g ˚C)

....

.... [3 marks]

1 a What is the difference between a batch process and a continuous process?

....

.... [2 marks]

b Why are pharmaceuticals made in batch processes?

....

.... [2 marks]

c Why are some chemicals, like ammonia, made using continuous processes?

.... [2 marks]

2 a List four things that affect the cost of making and developing a pharmaceutical drug.

....

....

....

.... [4 marks]

b Explain why it is often expensive to make and develop new pharmaceutical drugs.

....

....

.... [3 marks]

3 a Explain why it is important to manufacture pharmaceutical drugs that are as pure as possible.

.... [1 mark]

b Anya and Mark are researchers who have returned with a newly discovered plant which they think may contain an important chemical **Z**. Describe how they could extract the chemical from the plant and compare it to two other chemicals, **P** and **R**.

....

....

.... [3 marks]

c Anya and Mark test chemical **Z** for purity. They produce these results.

Chemical	Melting point °C	Boiling point °C	Chromatographic shift Rf
Z (new)	63.5 °C	124 °C	2 cm in 10 mins
P (pure old)	64.1 °C	122 °C	2 cm in 10 mins
R (pure old)	63.0 °C	126 °C	1.5 cm in 10 mins

Explain why they think that chemical **Z** may be a sample of impure **P** and not **R**.

....

....

.... [3 marks]

1 a Explain why diamond, graphite and fullerenes are allotropes of carbon.

.... [1 mark]

b Label these carbon structures.

.... [3 marks]

c Explain why diamond, graphite and fullerenes are allotropes of carbon.

.... [1 mark]

2 a Why can nanotubes be useful in tennis rackets?

.... [1 mark]

b Why can nanotubes be used in electrical circuits?

.... [1 mark]

c Explain why fullerenes can be used in new drug delivery systems.

.... [1 mark]

3 a List four physical properties of diamond.

....
....
....
.... [4 marks]

b List four physical properties of graphite

....
....
....
.... [4 marks]

c Explain why diamond is used in cutting tools and jewellery. Use ideas about properties.

....
.... [2 marks]

d Explain why graphite is used in pencil leads and lubricants. Use ideas about properties.

....
.... [2 marks]

e Explain why diamond and graphite have a giant molecular structure.

.... [2 marks]

G–E

D–C

G–E

D–C

G–E

D–C

Tom and Leah compare the energy per gram from two fuels, A and B. Describe the experiments they carry out, including a diagram of the apparatus they use, and explain the calculations they need to do.

❗ The quality of written communication will be assessed in your answer to this question.

[6 marks]

C3 Chemical economics

You can refer to the periodic table (page 248) to answer the questions in this section.

1 a Explain why Co is an element but CO is a compound.

.... [1 mark]

b Use the words to complete the sentences.

electrons   large   negative   neutral   neutrons   positive   protons   small

An atom has a nucleus surrounded by .... A nucleus has a charge that is

.... The surrounding .... have charges that are

.... An atom has a charge that is .... and has

a very .... mass.   [5 marks]

c Finish the table to show the relative mass and charge of the particles of an atom.

	Relative charge	Relative mass
electron		0.0005 (zero)
proton	+1	
neutron		

[2 marks]

D–C

2 a What is meant by the atomic number?

.... [1 mark]

b What is meant by the mass number?

.... [1 mark]

G–E

c What is the total number of protons, electrons and neutrons in an atom of potassium? Complete the table.

Atomic number	Mass number	Number of protons	Number of electrons	Number of neutrons
19	39			

[3 marks]

D–C

3 a What is an isotope? .... [1 mark]

b Isotopes of an element have different numbers of neutrons in their atoms. Complete the table.

Isotope	Electrons	Protons	Neutrons
$^{12}_{6}C$	6		
$^{14}_{6}C$			

[2 marks]

D–C

4 a How many shells do the electrons in an atom of magnesium occupy?

.... [1 mark]

G–E

b Describe how the elements are arranged in the periodic table.

....

.... [3 marks]

c Draw the electronic structure for the element aluminium. Explain why **three** shells are needed for its electrons.

....

.... [3 marks]

D–C

d Which element has the electronic structure 2.8.7 .... [1 mark]

C4 The periodic table   215

1 a What is an ion?

.... [1 mark]

b Put these symbols into the correct column in the table.

H   H$^+$ O^{2-} H$_2$O   Mg   CO$_2$ Cl   Cl$^-$ Cl$_2$ Na$^+$ Br$_2$ SO$_4^{2-}$ F$_2$

Atom	Ion	Molecule

[3 marks]

c Put **M** next to the box that describes a **metal atom**, and **N** next to the box that describes a **non-metal**.

An atom that has extra electrons in its outer shell and needs to **lose** them to be stable. ☐

An atom that has 'spaces' in its outer shell and needs to **gain** them to be stable. ☐   [1 mark]

d Draw a diagram to show how an atom of lithium transfers an electron to an atom of fluorine.

[2 marks]

e Explain why lithium forms a **positive** ion.

.... [1 mark]

f Explain why fluorine forms a **negative** ion.

.... [1 mark]

g Explain how the ionic bonding in lithium fluoride means that it becomes a solid.

....

.... [2 marks]

h What is the formula of calcium chloride if the calcium ion is Ca^{2+} and the chloride ion is Cl$^-$?

.... [1 mark]

2 a Fill in the table with a tick (✓) or cross (✗) to denote whether sodium chloride in these different forms conducts electricity or not.

solid		liquid		solution	

b Which has the higher melting point, sodium chloride or magnesium oxide?

.... [1 mark]

c Describe the structure of magnesium oxide.

.... [1 mark]

d When is magnesium oxide able to conduct electricity?

.... [1 mark]

The periodic table and covalent bonding

1 a Which kind of bonding exists between metals and non-metals?

.... [1 mark]

b Carbon dioxide and water form from bonding between non-metals. Do they conduct electricity?

.... [1 mark]

c Non-metals combine by covalent bonding. What happens in covalent bonding?

.... [1 mark]

d How do you describe water and carbon dioxide? Underline the correct answer.

   giant ionic lattice   simple molecules with weak intermolecular force

   simple molecules with strong intermolecular force   small ionic lattices   [1 mark]

2 a What is a group on the periodic table? Give three elements in group 6.

....

.... [2 marks]

b What is a period on the periodic table? Give three elements in period 2.

....

.... [2 marks]

3 a Explain why magnesium belongs to group 2 in the periodic table.

.... [1 mark]

b Explain why fluorine belongs to the same group as chlorine in the periodic table.

.... [1 mark]

c Explain why neon is in period 2 but potassium is in period 4 in the periodic table.

....

.... [2 marks]

d In which group is boron?

.... [1 mark]

e In which period is phosphorus?

.... [1 mark]

4 a From 1829 to 1869 Newlands and Mendeleev contributed to the main stages in the classification of elements. Who was the third scientist and what was the nature of his evidence?

....

.... [2 mark]

b What did both Newlands and Mendeleev notice about the behaviour of elements that helped them to start to classify them?

.... [1 mark]

1 a Write down the names of three group 1 metals.

.... [1 mark]

b Why are group 1 metals stored under oil?

.... [1 mark]

c How do group 1 metals react with water?

....

....

.... [3 marks]

d Why are the group 1 elements called **alkali metals**?

.... [1 mark]

e Construct a word equation for the reaction between potassium and water.

.... [1 mark]

f Predict what will happen when rubidium is added to water.

....

....

.... [3 marks]

g Which of rubidium or caesium will react more vigorously? Explain your answer.

....

.... [2 marks]

2 Explain why group 1 elements have similar properties.

.... [1 mark]

3 a Akira and Joe carry out some flame tests with compounds of group 1 metals. They find the colours of the compounds of two of these metals. Write down the colours they find.

metal in compound	Sodium	Potassium
colour of flame		

[2 marks]

b Joe found a bottle without a label. He found that the compound in the bottle gave a red flame. Which group 1 metal was in the compound?

.... [1 mark]

c Akira and Joe have been asked to demonstrate to a junior class how group 1 chemicals are used in fireworks to give different colours. How will they demonstrate this to show three different flames? You may use a diagram to help explain your answer.

....

....

....

.... [4 marks]

1 a What are the elements chlorine, bromine and iodine called?

.... [1 mark]

b What is the name of another element in this group? .... [1 mark]

c Give two uses for chlorine.

....

.... [2 marks]

d Give one use for iodine. .... [1 mark]

e Write in the table the physical appearance of the group 7 elements at room temperature.

	Physical appearance
chlorine	
bromine	
iodine	

[2 marks]

2 a Lithium reacts with iodine to make lithium iodide. Construct the word equation for this reaction.

.... [1 mark]

b Construct the word equation for the reaction between sodium and bromine.

.... [1 mark]

c Construct the balanced symbol equation for the reaction between potassium (K) and chlorine.

.... [2 marks]

3 a Describe the reactivity of the elements in group 7 either up or down the group.

.... [1 mark]

b Iodine and sodium bromide are made when bromine solution is bubbled through sodium iodide. Construct the word equation for this reaction.

.... [1 mark]

c What would you see if chlorine was bubbled through potassium iodide?

.... [1 mark]

d What type of reaction is this? .... [1 mark]

e Construct the word equation for the reaction between potassium iodide and chlorine.

.... [1 mark]

f Construct the balanced symbol equation for the reaction between potassium iodide (KI) and chlorine.

.... [1 mark]

1 a Which of these elements is a transition metal? Put a ring around the correct name.

beryllium   vanadium   germanium   selenium   [1 mark]

b What are the symbols for copper, iron, and the transition metal in question 1 a?

.... [3 marks]

c A compound that contains a transition element is often coloured. Match the colours to the compounds.

copper compounds		often pale green
iron(II) compounds		often orange/brown
iron(III) compounds		often blue

[1 mark]

d Transition elements or their compounds are often catalysts. What is a catalyst?

.... [1 mark]

e Give two examples of transition elements as catalysts.

....

.... [2 marks]

2 a What is meant by thermal decomposition?

.... [1 mark]

b Zinc oxide and carbon dioxide are made in the thermal decomposition of zinc carbonate.

i Construct the word equation for this reaction.

.... [1 mark]

ii How would you test for the carbon dioxide given off?

.... [2 marks]

c What two substances are made when manganese carbonate is heated?

.... [1 mark]

d Construct the word equation for the thermal decomposition of copper carbonate.

.... [1 mark]

e What would you see when copper carbonate decomposes?

.... [1 mark]

3 a If two solutions are added together and an insoluble solid is formed, what is the insoluble solid called?

distillate   filtrate   precipitate   .... [1 mark]

b Describe how you would use a solution of sodium hydroxide to identify the presence of the transition metal ions: Cu^{2+} Fe^{2+} Fe^{3+}

Describe what you would do and what you would **see**.

....

....

....

.... [5 marks]

1 a Metals have specific properties that make them useful. Which properties are necessary for making jewellery, bridges and saucepans?

....

.... [3 marks]

G–E

b Metals have specific properties that make them suitable for different uses.

Property	Iron	Copper	Aluminium	Tin	Gold	Silver	Lead
hardness in mohs	4	3	2.8	1.5	2.5	2.5	1.5
density in g/cm³	7.9	8.9	2.7	7.3	19.2	10.5	11.3
electrical conductivity × 10⁷ in siemen/cm	1.00	5.96	3.50	0.917	4.52	6.30	0.455
melting point in °C	1080	1358	933	505	1338	1234	601

D–C

i Explain why you would make electrical cables out of silver or copper rather than lead or iron.

.... [1 mark]

ii Choose which metal you would use to make a model aircraft. Justify your answer.

....

.... [2 marks]

iii What property of tin makes it more useful in solder than copper?

.... [1 mark]

2 a

electrons from outer shells of metal atoms are free to migrate

metal ions

How are particles in a metal held together?

....

....

.... [1 mark]

G–E

b Explain why metals have high melting points and high boiling points.

....

....

.... [1 mark]

D–C

3 a Under what conditions can some metals become superconductors?

.... [1 mark]

G–E

b Describe what is meant by a superconductor.

.... [1 mark]

c What are the potential benefits of superconductors?

....

....

.... [3 marks]

D–C

1 a Explain why water is an important resource for many industrial chemical processes.

.... [1 mark]

b Reservoirs are important water resources in the UK. Name three other important water resources.

.... [3 marks]

c Each region that these reservoirs serve needs 1100 units of water.

Reservoir	A	B	C	D
Volume of reserves (units)	1000	780	650	2400

i Which region has surplus supplies? .... [1 mark]

ii By how much is the shortfall of water needed in region B? .... [1 mark]

d Pollutants and other substances can be present in water before it is purified. Name three.

.... [3 marks]

e The water in a river is cloudy and often not fit to drink. To turn it into the clean water in taps it is passed through a **water purification** works.

[   ]   [   ]   [   ]

Label the three main stages in water purification and explain the three processes in detail.

....
....
.... [4 marks]

f How do some pollutants get into the water before or after purification?

.... [2 marks]

2 a Silver nitrate reacts with potassium iodide to produce potassium nitrate and silver iodide. Construct the word equation for this reaction.

.... [1 mark]

b Silver nitrate is used to test for halide ions in water. What colour precipitates are made?

Test reagent	Chloride ions	Bromide ions	Iodide ions
silver nitrate			

[3 marks]

c Three samples of water **A**, **B** and **C**, have the following reactions with test reagents. What type of sodium salts do the samples contain? Complete the boxes beneath the results table to identify the sodium salt in **A**, **B** and **C**.

Test reagent	A	B	C
silver nitrate	white solid	yellow solid	no solid
barium chloride	no solid	no solid	white solid
sodium salt contained			

[3 marks]

d Construct the word equation of a reaction between silver nitrate and a sodium salt that will produce a cream solid.

.... [2 marks]

C4 The periodic table

Jo and Sam have been given six salts **A**, **B**, **C**, **D**, **E** and **F** that need identifying. They know they contain Na^+, K^+, Cu^{2+}, Fe^{2+}, Li^+ and Fe^{3+} ions.

Describe the two experiments that they can do and explain how they would identify which ion each salt contained.

❗ The quality of written communication will be assessed in your answer to this question.

....
....
....
....
....
....
....
....
....
....
....
....
....
....
....
....
....
....
....
....
....
....
....
....
....
....
....
....
....
....
....
....

[6 marks]

1 a Melissa and Eve are running in a 1500 m race. Melissa finishes in 300 s. Eve takes 250 s.

  i Who runs faster, Melissa or Eve?

  .... [1 mark]

  ii How do you know?

  .... [1 mark]

  b Eve's father got a speeding ticket. His car was photographed twice as it crossed white lines in the road 1.5 m apart. He passed over eight lines in the 0.5 s between each photograph. How fast was he travelling?

  ....

  ....

  .... [3 marks]

2 Freddie is on holiday in France. He travels 390 km in 3 hours on a French motorway.

  a Calculate the average speed of his car on this journey in km/h.

  ....

  .... [2 marks]

  b Why is your answer the average speed of the car?

  .... [1 mark]

  c The speed limit on the motorway is 130 km/h. Did the car break the speed limit? Explain your answer.

  ....

  .... [1 mark]

3 The graph shows Ashna's walk to her local shop and home again.

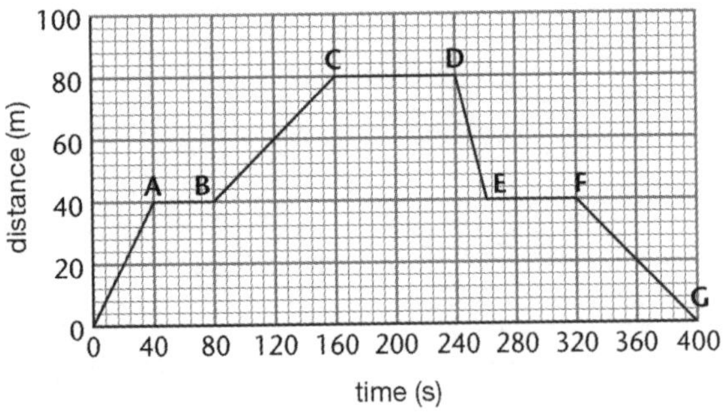

  a Between which points did Ashna wait at traffic lights to cross the road?

  .... [1 mark]

  b How long did she spend in the shop?

  .... [1 mark]

  c How far is the shop from her home?

  .... [1 mark]

  c During which part of her journey did she walk fastest?

  .... [1 mark]

1 Darren is riding his bicycle along a road. The speed–time graph shows how his speed changed during the first minute of his journey.

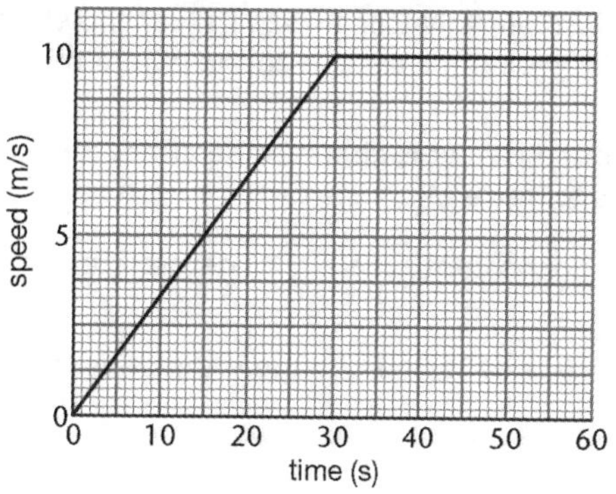

a Describe how the speed changed, giving as much detail as possible.

....

....

....

.... [4 marks]

b Darren makes the same journey the next day but:
  – increases his speed at a steady rate for the first 20 s, reaching a speed of 10 m/s
  – travels at a constant speed for 10 s
  – slows down at a steady rate for 15 s to a speed of 5 m/s
  – travels at a constant speed of 5 m/s.

Plot the graph of this journey on the same axes. [4 marks]

c How could you calculate the distance Darren travelled in the first minute of his original journey?

.... [1 mark]

2 The way in which the speed of a car changes over a 60 s period is shown in the table.

a Plot a speed–time graph for the car using the axes given.

time in s	speed in m/s
0	0
5	5
10	10
15	15
20	15
25	15
30	15
35	15
40	15
45	11
50	7
55	3
60	0

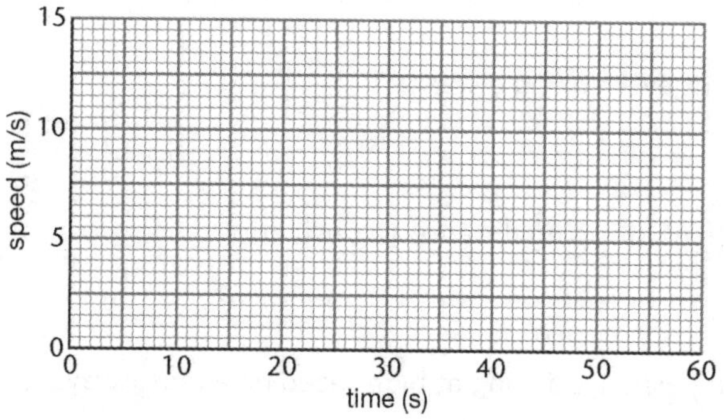

[4 marks]

b Calculate the deceleration between 40 and 55 seconds.

....

....

....

.... [3 marks]

1 The engine has just been switched on in the car shown in the diagram. P is the forward force on the car due to the pull of the engine.

a Put a (ring) around the word that best describes the motion of the car as it starts to move.

accelerate   constant speed   decelerate   **[1 mark]**

b i How will the size of force P change if the driver presses the accelerator pedal harder?

.... **[1 mark]**

ii How will the motion of the car change?

.... **[1 mark]**

c On another day the car carries four people and a fully loaded boot so its mass increases.

How will its motion change if the value of P is the same as in a?

.... **[1 mark]**

---

2 A car of mass 500 kg accelerates steadily from 0 to 40 m/s in 20 s.

a What is its acceleration?

....

.... **[2 marks]**

b What is the size of the resultant force which produces this acceleration?

....

.... **[2 marks]**

---

3 Helen is driving her car on a busy road when the car in front brakes suddenly. She puts her foot firmly on her brake pedal and just manages to stop without hitting the car in front.

a What is meant by 'braking distance'?

.... **[1 mark]**

b What is meant by 'thinking distance'?

.... **[1 mark]**

c How can Helen's stopping distance be calculated?

.... **[1 mark]**

d Later that day Helen is driving at high speed on a motorway.

i How will Helen's thinking distance change?

.... **[1 mark]**

ii Explain why.

.... **[2 marks]**

e Write down two things, apart from speed, that could increase a driver's thinking distance.

....

.... **[2 marks]**

Work and power

1 Hilary lifts a parcel of weight 80 N onto a shelf 2 m above the ground.

a Why would she do more work if the parcel weighed 120 N?

.... [1 mark]

b Why would she do less work if she lifted the parcel onto a shelf 1.6 m high?

.... [1 mark]

c Hilary tries to lift a parcel weighing 500 N but she cannot move it. How much work does she do?

.... [1 mark]

d Calculate the amount of work Hilary does when she lifts the 80 N parcel onto the shelf.

....

.... [2 marks]

2 Chris and Abi both have a mass of 60 kg. They both run up a flight of stairs 3 m high.
Chris takes 8 s and Abi takes 12 s.

a What can you say about the amount of work that each does?

.... [1 mark]

b What can you say about the power of Chris and Abi?

.... [1 mark]

c Calculate Chris' weight. [Take g = 10 N/kg.] ....

.... [2 marks]

d How much work does Chris do in running up the stairs? ....

.... [2 marks]

e Calculate Chris' power. ....

.... [2 marks]

f Calculate Abi's power.

....

.... [2 marks]

3 The table shows the fuel consumption of three cars in miles per gallon (mpg).

car fuel	consumption (mpg)
A	48
B	34
C	28

a Which car has the best fuel consumption?

.... [1 mark]

b Which car is likely to be most powerful?

.... [1 mark]

c We should keep our fuel consumption to a minimum to protect the environment. Why?

....

.... [3 marks]

Energy on the move

1 a What is meant by kinetic energy?

.... [1 mark]

b Look at the table. Which possesses more kinetic energy?

	mass in kg	speed in m/s		mass in kg	speed in m/s
A	2	5	**B**	2	7
C	20	2	**D**	15	2

i A or B   .... [1 mark]

ii C or D   .... [1 mark]

2 a Use the data about the fuel consumption of petrol and diesel cars to answer the following questions.

engine size in litres	fuel consumption in mpg	
	petrol	**diesel**
1.6	44	60
2.0	40	51

Which car has the best fuel consumption?

.... [1 mark]

b Use the data about fuel consumption to answer the following questions.

car	fuel	engine size in litres	miles per gallon	
			urban	**non-urban**
Renault Megane	petrol	2.0	25	32
Land Rover	petrol	4.2	14	24

i Suggest why fuel consumption is better in non-urban conditions.

....
.... [2 marks]

ii How many gallons of petrol would a Land Rover use on a non-urban journey of 96 miles?

.... [2 marks]

iii Suggest why a Renault Megane uses less fuel for the same journey?

.... [1 mark]

3 a How do battery-powered cars pollute the environment?

.... [1 mark]

b Give one advantage and one disadvantage of solar-powered cars compared to battery-powered cars.

....
.... [2 marks]

P3 Forces for transport

1 Kevin was involved in an accident on the M1 motorway. Luckily he was not seriously hurt but his car was badly damaged. Kevin's car had crumple zones, seat belts, airbags and ABS brakes.

a Which safety feature

i badly damaged his car but helped to protect Kevin?

.... [1 mark]

ii inflated and squashed to stop the steering wheel pushing into Kevin's chest?

.... [1 mark]

iii stopped Kevin being thrown through the windscreen?

.... [1 mark]

b Finish the table by describing how each feature helps to reduce Kevin's injuries.

safety feature	how it works
seat belt	
crumple zones	
air bag	

[3 marks]

c Explain why seat belts must always be replaced after a car crash.

.... [1 mark]

d Some safety features are intended to prevent accidents. Some are intended to protect occupants in the event of an accident. Give two examples of safety features that are intended to prevent accidents.

....

.... [2 marks]

2 Marie was a passenger in a car travelling at 25 m/s when the driver braked sharply to avoid a dog that had run into the road. Marie was wearing her seat belt and this brought her to a stop in 0.5 s. Marie's mass is 55 kg.

a Calculate the average force the seat belt exerted on her body.

....

....

.... [2 marks]

b If Marie had not been wearing her seat belt, she would have hit the windscreen which would have brought her head to a stop in 0.002 s. Calculate what the average force on her head would have been.

....

....

.... [2 marks]

G–E

D–C

G–E

D–C

1 Charlie drops a golf ball and a ping pong ball from a height of 30 cm above the ground.

Both balls hit the ground together although their masses are different.

Why do they both hit the ground at the same time?

.... [1 mark]

2 Sarah is a sky diver. She has a mass of 60 kg.

a On the diagram mark and name the forces acting on Sarah as she falls. [2 marks]

b As Sarah leaves the aircraft, she starts to accelerate.

Describe any difference in the size of the forces acting on her just after leaving the aircraft.

.... [1 mark]

c Sarah's acceleration decreases as she falls. Explain why.

.... [1 mark]

d i Eventually she is travelling at a constant speed. What is this speed called?

.... [1 mark]

ii Describe any difference in the size of the forces acting on her when she is falling at a constant speed.

.... [1 mark]

3 Racing cyclists try to streamline their shape.

a Why do they do this?

....

.... [2 marks]

b Give two ways by which they do this.

....

....

.... [2 marks]

1 Finish the sentences. Choose words from this list.

    gravitational potential energy (GPE)   kinetic energy (KE)

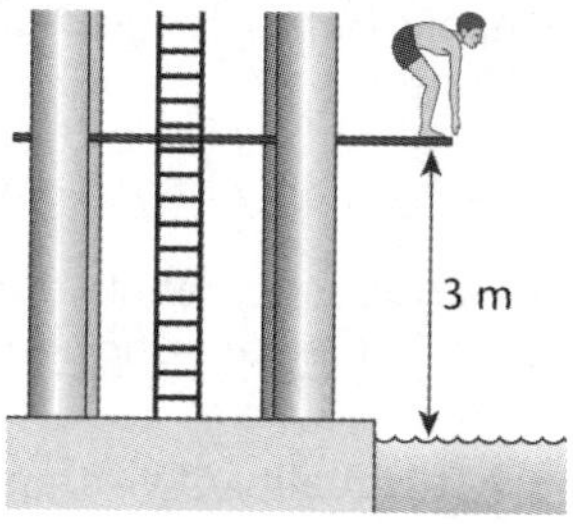

Rob is about to dive into the swimming pool.

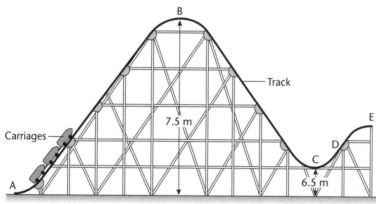

3 m

He has .... As he dives ....

changes to .... Rob climbs to the 10 m board.

He has more .... than before.

[4 marks]

---

2 The diagram shows a roller coaster. The carriages are pulled up to B by a motor and then released.

Track

B

7.5 m

E

D

Carriages

C

6.5 m

A

a At which point, A, B, C, D or E, do the carriages have the greatest gravitational potential energy?

.... [1 mark]

b How does the gravitational potential energy at this point change when the carriages are full of people?

.... [1 mark]

c At which point, A, B, C, D or E, do the carriages have the greatest kinetic energy?

.... [1 mark]

d Describe the main energy change as the carriages move from B towards C.

.... [1 mark]

e Why must the height of the next peak at E be less than that at B?

....

.... [1 mark]

f The theme park decides to build a faster roller coaster. Suggest how they could modify the design to achieve this.

....

....

....

.... [3 marks]

Thanks to the design of modern cars, there are fewer serious casualties today than there were in the past.

The occupants from many crashes now walk away from the scene, unharmed.

Describe the safety features that modern cars have to reduce injury in the event of an accident and explain the scientific principles behind them.

❗ The quality of written communication will be assessed in your answer to this question.

....
....
....
....
....
....
....
....
....
....
....
....
....
....
....
....
....
....
....
....
....
....
....
....
....
....
....
....

[6 marks]

1 Liz rubs a polythene strip with a cloth. The strip becomes charged. It can pick up small pieces of paper.

   a What charge does the polythene strip gain?

   .... [1 mark]

   b i What material could she rub to obtain a strip with a different charge?

   .... [1 mark]

   ii What effect will this strip have on the small pieces of paper?

   .... [1 mark]

   c Craig rubs a copper strip with a cloth. Explain why it has no effect on the small pieces of paper.

   ....
   .... [2 marks]

G–E

2 Sally works in a factory where computers are manufactured. She wears special clothing while she is working in the factory.

   a Why does Sally have to wear special clothing in the factory?

   ....
   .... [1 mark]

   b i Sally wears a wristband which is connected via a wire to the ground. Why is this important?

   ....
   .... [1 mark]

   ii The wristband has a high value resistor inside it. Why is this necessary?

   ....
   .... [1 mark]

D–C

3 Use your knowledge of electrostatics to explain the following:

   a You can get a shock from touching a tap after walking on a nylon carpet.

   ....
   .... [2 marks]

G–E

   b Synthetic clothing sticks together when it comes from the tumble dryer.

   ....
   .... [2 marks]

   c Dust sticks to television screens.

   ....
   .... [1 mark]

D–C

1 a Why is an electrostatic precipitator with a charged plate placed in power station and factory chimneys?

.... [1 mark]

b What type(s) of power station need an electrostatic precipitator?

.... [1 mark]

c An electrostatic precipitator contains some wires and plates which are connected to a high voltage supply.

If the wires are given a negative charge, what charge will be gained by the plates?

.... [1 mark]

d What charge do the soot particles gain as they pass close to the wires?

.... [1 mark]

2 In a coffee factory the jars are being filled with coffee from a large funnel. Lots of coffee spills over the sides of the jar.

a Why does the coffee spill over the sides?

.... [1 mark]

b What can be done to prevent wasting so much coffee?

....
.... [2 marks]

3 A defibrillator delivers an electric shock through the chest wall to the heart.

a What is the purpose of defibrillation?

.... [1 mark]

b What does the electric shock do to the heart?

.... [1 mark]

c The paddles of a defibrillator, charged from a high voltage supply, are placed on the patient's chest. Why should the chest be clean shaven and dry?

....
.... [2 marks]

d Why does the operator call out 'clear' before using the paddles?

....
.... [1 mark]

e A current of about 50 A passes through the patient for about 4 ms (0.004 s).

In general, such a large current would be fatal. Why can it be used in this situation?

....
.... [1 mark]

P4 Radiation for life

1 Adam sets up the circuit shown. The lamp does not light.

a i Why doesn't the lamp light?

.... [1 mark]

ii Change Adam's circuit so that the lamp will light.

[1 mark]

G–E

b Adeela sets up the circuit shown. She then moves the variable resistor to include a greater length of wire into the circuit.

i What effect will this have on the brightness of the lamp?

.... [1 mark]

variable resistor

ii Add an ammeter and voltmeter to Adeela's circuit to allow her to measure the current through the lamp and the voltage across the lamp. [2 marks]

iii If the voltmeter reads 6 V and the ammeter 0.25 A, calculate the resistance of the lamp.

....

.... [3 marks]

D–C

2 **a i** Rob is in the garden using a lawnmower. Why is it important that he has a circuit breaker in the circuit while using the lawnmower?

.... [1 mark]

ii What is the advantage of a circuit breaker over a fuse?

.... [1 mark]

G–E

b Rob's son has a toy lawnmower which operates with batteries. Give **two** differences between the voltage from batteries and mains voltage.

....

.... [2 marks]

D–C

3 **a i** Give the functions of the live (L), neutral (N) and earth (E) wires in the plug shown. [3 marks]

ii Which wire, live, neutral or earth, is a safety wire?

.... [1 mark]

iii Why is the safety wire coloured differently to the other wires?

....

.... [2 marks]

b i Mike was boiling a kettle when the fuse blew. He was going to replace the fuse with a small iron nail. Why is this not a good idea?

.... [1 mark]

ii Why are fuses always connected in the live wire?

.... [2 marks]

D–C

iii The kettle is made from metal. How will the fuse and earth wire stop Mike receiving an electric shock if he touches the kettle when it is faulty?

....

.... [4 marks]

P4 Radiation for life 235

1

G–E

a Why is ultrasound unable to travel in a vacuum?

.... [1 mark]

b The distance covered by four compressions and four rarefactions is 6 mm. What is the wavelength of the ultrasound?

.... [1 mark]

c Ultrasound is used by dolphins and whales to communicate. Why do animals under the water detect the ultrasound before those on the surface?

....

.... [2 marks]

d Ultrasound can be used to clean teeth. What property of ultrasound makes it useful for this?

.... [1 mark]

D–C

e A device giving off ultrasound can be used by the police to measure the speed of a moving car. Why don't the motorists realise their speed is being measured?

.... [1 mark]

f Light is an example of a different type of wave. What is the difference between the two types of wave?

....

.... [2 marks]

2 High-powered ultrasound is used to treat a patient with kidney stones.

a How does ultrasound do this?

....

....

.... [3 marks]

D–C

b Why must high-powered ultrasound be used?

....

.... [2 marks]

3

a Why does a pregnant woman usually have an ultrasound scan?

.... [1 mark]

b Give two other uses of ultrasound scanning.

....

.... [2 marks]

G–E

c The ultrasound waves used for foetal scanning vibrate 1 000 000 times a second. What is the frequency of the ultrasound?

.... [1 mark]

P4 Radiation for life

1 Jasmine is going to measure the activity of a radioactive alpha source.

a What is meant by the 'activity' of a radioactive source?

.... [2 marks]

b i The table below shows the results of Jasmine's experiment.

time in seconds	source	count
30	none	6
30	none	8
30	none	10
30	alpha	798
30	alpha	758
30	alpha	718

What is the activity of the source?
Show how you work out your answer.

....

....

....

.... [4 marks]

ii Would you expect the activity of the source to increase or decrease as time passes?

.... [1 mark]

iii Why did Jasmine repeat her readings several times?

.... [2 marks]

G–E

2 a Complete the table showing the three types of nuclear radiation.

type of radiation	charge	what it is	particle or wave
alpha			
beta			
gamma			

[3 marks]

b Name the type of nuclear radiation that:

i is the most penetrating .... [1 mark]

ii is stopped by several sheets of paper .... [1 mark]

iii has the greatest mass .... [1 mark]

iv does not change the composition of the nucleus .... [1 mark]

v travels at about one-tenth the speed of light .... [1 mark]

D–C

3 a What is meant by 'half-life'? .... [1 mark]

b

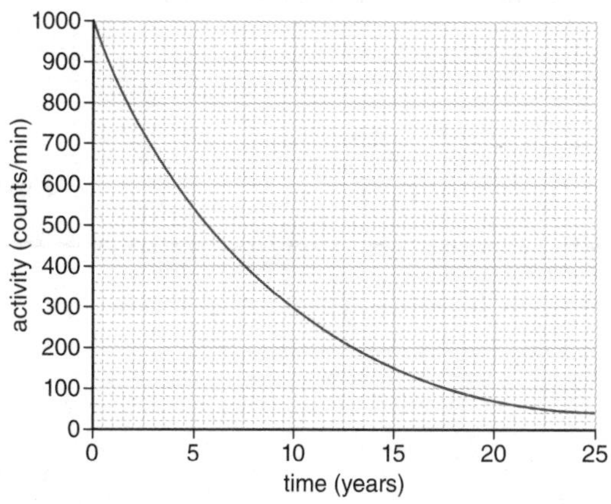

The graph shows how the activity of cobalt-60 changes with time. Use the graph to find the half-life of cobalt-60.

Show clearly, on the graph, how you got your answer. [1 mark]

D–C

1 a What is meant by background radiation?

.... [1 mark]

b Why should airline crew be made aware of radiation risks?

....

.... [2 marks]

2 The Environment Agency is worried that water in a river is being polluted by a nearby factory. Explain how it could use a radioactive tracer to investigate this.

....

....

....

.... [2 marks]

3 Smoke alarms use a source of alpha radiation in a small chamber.

oppositely charged plates

americium-241 source of alpha particles

a Why is alpha radiation more suitable than either gamma or beta radiation for use in a smoke alarm?

....

.... [1 mark]

b Explain how the smoke alarm works.

....

....

....

.... [4 marks]

4 An archaeological dig in Italy has uncovered some ancient skeletal remains. Explain how scientists can help to identify the period of their origin.

....

....

.... [2 marks]

1 a Ben has injured his leg playing rugby and the doctor wants to find out what exactly the problem is. How will the doctor diagnose the problem with Ben's leg?

....

.... [1 mark]

b If the doctor needs to operate on Ben he will need to use medical instruments that have been treated by gamma radiation. Why?

....

....

.... [2 marks]

G–E

c Ben has a metal plate in his other leg from a previous injury. How can the doctor see where the plate is?

....

.... [2 marks]

d Who will be in the room with Ben when he is having the test carried out?

....

.... [1 mark]

D–C

2 a James is being treated for lung cancer. Why is nuclear radiation suitable for treating cancer?

.... [1 mark]

b How is the cobalt-60, which is used for the treatment, prepared?

.... [1 mark]

c Why is the cobalt-60 transported in a lead-lined box to the hospital?

.... [2 marks]

G–E

3 a Heather has a problem with her breathing and the doctor wants to carry out some investigations using a radioactive tracer. What is a tracer?

....

.... [2 marks]

b Why is it used?

....

....

.... [2 marks]

c What sort of radiation should a tracer emit?

.... [1 mark]

D–C

1 A power station makes electricity.

G–E

a Label the diagram.

generator, source of energy, steam, water, turbine.

hot gas

cool gas

[5 marks]

b Use these words to complete the following sentences explaining how a power station works.

D–C

generator   **source of energy**   **steam**   **turbine**   **water**

The .... provides heat to boil the .... to

produce .... The pressure of the .... turns the

.... which turns the .... making electricity.   [6 marks]

2 In March 2011 an earthquake in Japan caused major problems for the nuclear power stations. Scientists were worried that the nuclear fuel may be released into the environment.

a Why would this be a problem?

.... [2 marks]

b If the reaction in the power station was not controlled what could happen?

.... [2 marks]

D–C

c i The electric pumps that brought water into the reactor had no electrical supply and so stopped working. Why was this dangerous?

.... [1 mark]

ii Eventually the reactor was flooded with sea water as a last resort. Why was this only used as a last resort?

.... [2 marks]

3 a What is nuclear fusion?

.... [1 mark]

G–E

b Why do some scientists dispute other scientists' claims about fusion?

.... [1 mark]

c Why is fusion difficult to achieve in a laboratory?

....
.... [2 marks]

D–C

d Why is fusion research a joint venture between different countries?

.... [1 mark]

P4 Extended response question

Many houses are now fitted with smoke alarms to warn of a fire. Some people are anxious about having a smoke alarm fitted because they have heard they contain dangerous nuclear radiation.

Describe how a smoke alarm works and explain, in as much detail as you can, why people who have them fitted would not be in danger from having one in the house.

❗ The quality of written communication will be assessed in your answer to this question.

....
....
....
....
....
....
....
....
....
....
....
....
....
....
....
....
....
....
....
....
....
....
....
....
....
....
....
....
....
....
....
....
....
.... **[6 marks]**

B1

I can explain why blood in arteries is under pressure.	
I can analyse data about heart disease in the UK.	
I can interpret simple data on diet.	
I can recall what is included in a balanced diet .	
I can describe how the human body defends itself against pathogens.	
I can interpret data which shows the incidence of disease around the world.	
I can name and locate the main parts of the eye.	
I can describe the nerve impulse as an electrical signal carried by nerve cells called neurones.	
I can state the general effects of each category of harmful drugs.	
I can describe the short-term and long-term effects of alcohol on the body.	
I can recall that the body maintains steady levels of temperature, water and carbon dioxide.	
I can describe how heat can be gained/ retained or lost by the body.	
I can recall that plant growth is controlled by chemicals called plant hormones.	
I can describe an experiment to show that plant shoots grow towards the light.	
I can recall that chromosomes are inside nuclei and that they carry information in the form of genes.	
I can recall that some human disorders such as cystic fibrosis and red-green colour blindness are inherited.	
I am working at grades G–E	

I can describe the factors that increase or decrease blood pressure.	
I can explain how a diet containing high levels of saturated fats and salt can increase the risk of heart disease.	
I can explain why protein deficiency is common in developing countries.	
I can calculate a person's BMI and their EAR for protein.	
I can explain the difference between active and passive immunity.	
I can describe changes in lifestyle that may reduce the risk of some cancers.	
I can describe the path taken by a spinal reflex.	
I can explain the differences between monocular and binocular vision.	
I can explain why damage to ciliated cells can lead to a 'smoker's cough'.	
I can interpret data on the alcohol content, measured in units of alcohol, of different alcoholic drinks.	
I can describe the dangerous effects of high and low temperatures on the body.	
I can explain how Type 2 diabetes can be controlled by diet but Type 1 diabetes needs insulin doses.	
I can describe shoots as being positively phototropic and negatively geotropic.	
I can describe the commercial uses of plant hormones.	
I can describe how sex chromosomes, XX in female and XY in male, determine the sex of an individual.	
I can explain the causes of genetic variation.	
I am working at grades D–C	

B2

I can use the characteristics of a organism to place it into a kingdom.	
I can recall the characteristics of the arthropod classes.	
I can explain the meaning of the term 'trophic level'.	
I can predict the effects of changes in numbers of one organism on a food chain.	
I can recall the types of organisms that act as decomposers.	
I can explain why nitrogen gas cannot be used directly by animals and plants.	
I can describe the resources that organisms compete for.	
I can describe one example of a relationship in which both organisms benefit.	
I can recognise some of the features of successful predators.	
I can describe how some animals are adapted to avoid being caught by predators.	
I can define the term 'evolution'.	
I can recall that the theory of natural selection was put forward by Charles Darwin.	
I can appreciate that an increase in human population will use more resources.	
I can describe some examples of pollution caused by an increase in human population.	
I can explain why some organisms have become extinct.	
I can describe how some endangered species can be conserved.	
I am working at grades G–E	

I can name organisms using the binomial system.	
I can recall the definition of a species.	
I can describe how energy is lost from food chains.	
I can explain why pyramids of energy and pyramids of biomass can be different shapes.	
I can describe how plants obtain nitrogen from the soil.	
I can describe the role of decomposers in the carbon and nitrogen cycles.	
I can explain the difference between parasitism and mutualism.	
I can explain the reasons for changes in predator and prey numbers.	
I can explain some of the adaptations that animals have for living in cold conditions.	
I can explain some of the adaptations that organisms have for living in dry conditions.	
I can use Darwin's theory of natural selection to explain how evolution occurs.	
I can state why Darwin's theory was not well accepted by many people.	
I can describe how indicator species can show levels of pollution.	
I can explain the causes and consequences of global warming.	
I can explain why it is considered important to set up conservation programmes.	
I can explain what sustainable development means.	
I am working at grades D–C	

C1

I know that LPG, petrol, diesel, paraffin, heating oil, fuel oils and bitumen are fractions from crude oil.	
I know that cracking converts large hydrocarbon molecules into smaller ones that are more useful, like petrol.	
I know energy value, availability, storage, cost, toxicity, pollution and ease of use are key when choosing fuels.	
I know combustion releases useful heat energy and complete combustion needs a good supply of oxygen.	
I know how photosynthesis, combustion and respiration affect the level of carbon dioxide in the air.	
I know the common pollutants found in air and where they come from, such as those from incomplete combustion.	
I can recognise that alkanes and alkenes are hydrocarbons and can recognise a hydrocarbon from its formula.	
I can work out the name of an addition polymer from the name of its monomer and the other way round.	
I can interpret information about the properties of polymers and their uses from data tables.	
I know many polymers are non-biodegradable and are disposed of in landfill sites or by burning or are recycled.	
I can explain that cooking is a chemical process as it cannot be reversed and a new substance is formed.	
I know antioxidants stop foods from reacting with oxygen and emulsifiers help water and oil to mix.	
I know that cosmetics can be synthetic or natural and esters are perfumes that can be made synthetically.	
I know that nail-varnish remover dissolves nail varnish colours and nail varnish colours are insoluble in water.	
I know the ingredients of paint are pigment, which gives colour; solvent which thins it; and binding medium.	
I know that thermochromic pigments change colour on heating, and phosphorescent pigment glows in the dark.	
I am working at grades G–E	

I can describe how fractional distillation separates crude oil in a column with a temperature gradient.	
I can interpret data about the supply and demand of crude oil fractions and describe the cracking of alkanes.	
I can suggest key factors that need to be considered when choosing fuels and interpret data to pick the best.	
I can describe an experiment to show the products of combustion of a hydrocarbon in a good supply of air.	
I can describe a simple carbon cycle involving the processes of photosynthesis, combustion and respiration.	
I can describe how the present-day atmosphere evolved from gases from inside the Earth and photosynthesis.	
I can recognise and interpret information on the displayed formulae of alkanes, alkenes and polymers.	
I can describe how bromine is used to test for an alkene, because orange bromine water decolourises.	
I can compare GORE-TEX® to nylon and say its breathable properties make it useful to active outdoor people.	
I can explain the environmental and economic issues related to uses and disposal of polymers.	
I can recall that protein molecules change shape when they are heated and that this is called denaturing.	
I can construct the balanced symbol equation for the decomposition of sodium hydrogencarbonate given some or all of the formulae.	
I can recall that alcohols react with acids to make an ester and water, and describe this as an experiment.	
I can explain why a perfume needs certain properties, e.g. being non-toxic and able to easily evaporate.	
I can describe paint as a colloid where particles are dispersed with particles of a liquid but are not dissolved.	
I can explain when thermochromic pigments are suitable and how phosphorescent pigments glow in the dark.	
I am working at grades D–C	

C2

I can describe the structure of the Earth as a sphere with a thin rocky crust, a mantle, and an iron core.	
I know that when igneous rock forms, big crystals are made if it cools slowly, and small crystals if it cools quickly.	
I know that granite, limestone, marble and aggregates are used in the construction of buildings and roads.	
I know that concrete is made from cement, sand, aggregate and water mixed together then allowed to set.	
I know that copper is extracted from its ore by heating the ore with carbon. The copper is then purified.	
I know that alloys are mixtures containing one or more metal elements. For example, brass, bronze and steel.	
I know that rusting needs iron, water and oxygen and that it is an example of oxidation, as oxygen is added.	
I know that aluminium is less dense than iron, that iron is magnetic and both are good electrical conductors.	
I know that in the Haber process ammonia is made from nitrogen (from the air) and hydrogen (from oil).	
I know the factors that affect the costs of making substances, such as price of energy, raw materials and wages.	
I know how universal indicator can estimate the pH of a solution, turning red at pH 1 and purple at pH 14.	
I know that an acid can be neutralised by a base or an alkali or that an alkali can be neutralised by an acid.	
I can know that nitrogen, phosphorus and potassium are three essential elements needed for plant growth.	
I know that fertilisers can be beneficial in increasing food supply but can cause problems like eutrophication.	
I know that in the electrolysis of concentrated sodium chloride solution chlorine and hydrogen are given off.	
I know that sodium chloride is an important raw material in the chemical industry as a source of chlorine.	
I am working at grades G–E	

I know the lithosphere is the Earth's crust and outer part of the mantle, and that it is made of tectonic plates.	
I know that the type of volcanic eruption depends on the type of magma with thick lava making it explosive.	
I know that limestone decomposes on heating to make calcium oxide and can write the word equation.	
I know that bricks are made from clay and cement is made when limestone and clay are heated together.	
I know how to purify copper by electrolysis and can explain the advantages and disadvantages of recycling it.	
I know that brass is an alloy of copper and zinc, that solder is made from lead and tin and can give some uses.	
I know that rusting is an oxidation reaction of iron and the word equation to make hydrated iron(III) oxide.	
I know that a car body made from aluminium will corrode less and will be lighter than one made from steel.	
I can write an equation for the Haber process and know that high pressure and an iron catalyst are needed.	
I know how factors affect costs in making substances, e.g. higher temperatures lead to higher energy costs.	
I know that acids neutralise bases to make salt and water and can show these changes in pH using indicators.	
I know how to make a salt and predict its name, e.g. sodium hydroxide and nitric acid makes sodium nitrate.	
I can identify the argument for the use of fertilisers as the world population is rising so more food is needed.	
I know that eutrophication and pollution of water supplies can result from the excessive use of fertilisers.	
I know that chlorine is made at the anode and hydrogen at the cathode in the electrolysis of brine.	
I can describe how sodium hydroxide and chlorine are used to make household bleach.	
I am working at grades D–C	

P1	
I can describe an experiment to measure the energy needed to change the temperature of a body.	
I can interpret data which shows there is no temperature change when materials melt, freeze or boil.	
I can describe energy-saving measures in the home.	
I can explain why trapped air is a very good insulator.	
I can identify the main features of a transverse wave and use the wave equation.	
I can identify the different types of electromagnetic waves and describe how they are reflected and refracted.	
I can describe how total internal reflection happens in an optical fibre.	
I can describe laser light and list some of its uses.	
I can recall the properties of infrared and microwave radiations that make them useful for heating things.	
I can recall that mobile phones use microwaves and that there are concerns about possible dangers.	
I can describe uses of infrared radiation.	
I can describe the differences between analogue and digital signals.	
I can describe how radiation is used for communication and the advantages of wireless technology.	
I can interpret information on analogue and digital signals.	
I can describe the effects of earthquakes.	
I can state the effects of exposure to ultraviolet radiation and recognise that sunscreen reduces damage.	
I am working at grades G–E	

I can interpret data on rate of cooling and understand the consequences of the direction of energy flow.	
I can understand the concepts of specific heat capacity and specific latent heat and use the equations.	
I can explain how energy is transferred and how losses to the atmosphere can be reduced.	
I can interpret data on energy-saving strategies and understand the importance of energy efficiency.	
I can describe the features of a transverse wave and determine wavelength and frequency from a diagram.	
I can understand that refraction is due to a wave travelling at different speeds in different materials.	
I can describe how light and the Morse code have been used for communication.	
I can describe how light behaves when its angle of incidence is below, above and equal to the critical angle.	
I can provide reasons for poor mobile phone reception.	
I can explain how scientists check each other's results by publishing their findings.	
I can describe how infrared signals carry information to control electrical and electronic devices.	
I can recall how the properties of digital signals led to the digital switchover for television and radio.	
I can describe how reflection and refraction of radio waves can be an advantage and can be a disadvantage.	
I can state the advantages and disadvantages of DAB broadcasts.	
I can recall the properties of P waves and S waves from an earthquake.	
I can interpret data about how sunscreen and skin tone can protect the skin from damage when sunbathing.	✓
I am working at grades D–C	

P2	
I can describe how photocells are used in remote locations to turn light into direct current electricity.	
I can describe how energy from the Sun can be harnessed by passive solar heating and wind turbines.	
I can describe how alternating current can be generated using the dynamo effect.	
I can describe the main stages in the production and distribution of electricity.	
I can identify greenhouse gases and describe how human activity leads to increased levels in the atmosphere.	
I can describe the difficulties in measuring global warming and why scientists need to work together on this.	
I can calculate the power rating of an appliance.	
I can recall that transformers can be used to increase or decrease voltage.	
I can recall the three types of radiation and give examples where these can be beneficial or harmful.	
I can describe how to handle radioactive materials safely.	
I can explain why stars are visible and some of the other items in our universe are not.	
I can compare the use of manned and unmanned craft to explore our Solar System.	
I can recall that asteroids are rocks and describe some of the consequences of Earth being hit by an asteroid.	
I can describe comets and near-Earth objects and how these may be observed.	
I can describe models of the Universe and how these have changed over time.	
I can recall that stars begin in clouds of gas and dust and have a finite lifetime.	
I am working at grades G–E	

I can describe advantages and disadvantages of using photocells to produce electricity.	
I can describe advantages and disadvantages of wind turbines.	
I can describe how a simple alternating current generator works and how to increase the output.	
I can calculate the efficiency of a power station.	
I can explain how human activity and natural phenomena both have effects on weather patterns.	
I can distinguish between opinion and evidence-based statements in the global warming debate.	
I can calculate the amount of electricity used in kilowatt hours and use this to find its cost.	
I can explain that transformers are used in the National Grid to reduce energy waste and costs.	
I can describe the relative penetrating power of alpha, beta and gamma radiations.	
I can describe some methods of disposing of radioactive waste.	
I can recall the relative sizes and nature of planets, stars, comets, meteors, galaxies and black holes.	
I can recall some difficulties of manned space travel and explain how information from space can be sent back to Earth.	
I can describe some of the evidence for past large asteroid collisions.	
I can describe how a collision between two planets could have resulted in the Earth-Moon system.	
I can recall that all galaxies are moving apart and that microwave radiation is received from all parts of the Universe.	
I can describe the end of the life cycle of small and large stars.	
I am working at grades D–C	

P1 and P2 Grade booster checklist

B3

I can identify the mitochondria in an animal cell.	
I can recall that DNA controls the production of proteins.	
I can recall some examples of different proteins.	
I can describe the function of enzymes.	
I can recall the word equation for aerobic respiration.	
I can explain why breathing rate and pulse rate increases during exercise.	
I can describe the main difference between unicellular and multicellular organisms.	
I can recall that sexual reproduction gametes join in fertilisation.	
I can describe the functions of red blood cells, white blood cells and platelets.	
I can recall the main types of blood vessels.	
I can describe the functions of the main parts of a plant cell.	
I can interpret data from a typical growth curve.	
I can describe the process of selective breeding.	
I can describe what is meant by genetic engineering.	
I can recall examples of natural and artificial clones.	
I can describe how to take a cutting.	
I am working at grades G–E	

I can explain why liver and muscle cells have large numbers of mitochondria.	
I can describe the shape of a DNA molecule.	
I can explain why enzymes are specific.	
I can describe the effect of temperature on enzyme action.	
I can recall the symbol equation for aerobic respiration.	
I can explain why anaerobic respiration occurs during exercise.	
I can explain some of the advantages of being multicellular.	
I can recall the uses of mitosis and meiosis in mammals.	
I can explain how a red blood cell is adapted to its function.	
I can label the main parts of the mammalian heart.	
I can identify simple differences between bacterial cells and plant and animal cells.	
I can recall the function of stem cells.	
I can recognise that selective breeding may lead to inbreeding.	
I can explain some potential benefits and risks of genetic engineering.	
I can describe some possible uses of cloning.	
I can describe what is meant by nuclear transfer.	
I am working at grades D–C	

B4

I can describe how to use collecting/counting methods.	
I can define biodiversity as the variety of different species living in a habitat.	
I can recall and use the word equation for photosynthesis.	
I can explain why plants grow faster in summer than in winter.	
I can recall that chlorophyll pigments in chloroplasts absorb light energy.	
I can recall the entry and exit points of materials involved in photosynthesis.	
I can describe diffusion.	
I can describe how carbon dioxide and oxygen diffuse in and out of plant leaves.	
I can describe how water travels through a plant.	
I can describe experiments to show how different conditions affect the transpiration rate.	
I can interpret data on NPK fertilisers.	
I can describe experiments to show the effects of mineral deficiencies.	
I can recall the key factors in the process of decay.	
I can recognise that food preservation techniques reduce the rate of decay.	
I can analyse data to show the effects of pesticides on food production.	
I can describe how intensive farming methods increase productivity.	
I am working at grades G–E	

I can calculate an estimate of a population size.	
I can compare the biodiversity of natural and artificial ecosystems.	
I can recall and use the balanced symbol equation for photosynthesis.	
I can explain why plants carry out respiration at all times.	
I can name and locate the main parts of a leaf.	
I can explain how leaves are adapted for photosynthesis.	
I can explain the net movement of particles in diffusion.	
I can describe the process of osmosis.	
I can recall that transpiration is the evaporation and diffusion of water from inside leaves.	
I can interpret data on transpiration rates.	
I can explain why plants need nitrates, phosphates, potassium and magnesium.	
I can relate mineral deficiencies to their symptoms.	
I can describe the effects of temperature, oxygen and water on the rate of decay.	
I can explain how food preservation methods reduce the rate of decay.	
I can describe how plants can be grown without soil (hydroponics).	
I can explain the advantages and disadvantages of biological control.	
I am working at grades D–C	

C3

I recognise that rusting is a slow reaction and that burning and explosions are very fast reactions.	
I can read values from a graph and compare rates of reaction by comparing gradients of graphs.	
I know particles collide in a chemical reaction and the higher the temperature the higher the rate of reaction.	
I know that a higher concentration or higher pressure increases the rate of reaction.	
I know that if a powdered reactant rather than a lump is used, the rate of reaction will be higher.	
I know that the rate of a reaction can be increased by adding a catalyst.	
I can calculate the relative formula mass of a substance from its formula, given the relative atomic masses.	
I understand that the principle of conservation of mass means that mass of reactants is the same as mass of products.	
I recognise that loss in filtration is one possible reason why percentage yield of a product is less than 100%.	
I understand that 100% atom economy means that all the atoms in the reactant go into the desired product.	
I know exothermic reactions release energy and endothermic reactions take in energy from the surroundings.	
I know a method for comparing the energy transferred in combustion by measuring temperature changes.	
I know that a continuous process carries on all of the time whereas a batch process starts and finishes.	
I know labour, energy, raw materials, research and testing affect costs of developing a pharmaceutical drug.	
I can recognise the structures of diamond, graphite and buckminsterfullerene, which are all forms of carbon.	
I know that diamond is lustrous, hard and does not conduct electricity but graphite conducts and is slippery.	
I am working at grades G–E	

I know that the rate of reaction measures how much product is formed in a fixed time in g/s or cm³/s.	
I know that the amount of product formed is directly proportional to the amount of limiting reactant used.	
I know that the rate of reaction depends on the number of collisions between reacting particles.	
I can draw sketch graphs to show the effects of changing temperature or concentration on the rate of reaction.	
I know that catalysts are specific to particular reactions and only small amounts are needed.	
I can draw sketch graphs to show the effects of surface area on the reaction rate and the amount of product formed.	
I can calculate the M_r of substances like $Ca(OH)_2$ that have formulae with brackets if given the A_r.	
I can show that mass is conserved during a reaction using a simple equation and the relative formula masses.	
I know and can use the formula: $\text{percentage yield} = \dfrac{\text{actual yield} \times 100}{\text{predicted yield}}$	
I know and can use the formula: $\text{atom economy} = \dfrac{M_r \text{ of desired products} \times 100}{\text{sum of } M_r \text{ of all products}}$	
I know that bond breaking is an endothermic process and bond making is an exothermic process.	
I know how to measure the energy given out by fuel, using a calorimeter, and can describe the experiment.	
I can explain why most pharmaceuticals are produced by batch and other bulk chemicals by a continuous process.	
I know that chemicals are extracted from plants by crushing, boiling, dissolving and using chromatography.	
I know that diamond, graphite and fullerenes are different structures of the same element, so are allotropes.	
I know that graphite can be used in lubricants, as it is slippery, and in pencil leads, as it is slippery and black.	
I am working at grades D–C	

C4

I know that an atom is neutral, a nucleus is positively charged and electrons are negatively charged.	
I know atomic numbers are the number of protons and mass numbers are numbers of protons and neutrons.	
I can recognise ions, atoms and molecules from formulae such as the ion Na^+, the atom C and molecule CO_2.	
I know that solid NaCl does not conduct electricity but that molten NaCl and NaCl solution both do.	
I know that there is ionic bonding between metals and non-metals and covalent bonding between non-metals.	
I know that a group of elements are those in a vertical column in the periodic table and with similar properties.	
I know that group 1 metals are known as the alkali metals, as when they react with water they form alkalis.	
I know that the compounds of lithium, sodium and potassium make red, yellow and lilac flames respectively.	
I know that fluorine, chlorine, bromine and iodine are the group 7 elements and are known as the halogens.	
I can construct the word equation for the reaction between a group 1 element and a group 7 element.	
I know the position of the transition elements in the periodic table and can deduce their names or symbols.	
I know thermal decomposition is a substance being broken down to at least two other substances by heat.	
I can list the physical properties of metals as being lustrous, hard and good conductors of heat and electricity.	
I can suggest properties needed by metals for given uses, such as kitchen pan metals need to be good heat conductors.	
I know that the water resources in the UK are lakes, rivers, aquifers and reservoirs and this water is purified.	
I know silver nitrate solution makes a white precipitate with chloride ions and a cream one with bromide ions.	
I am working at grades G–E	

I know that the nucleus is made of protons of mass 1 and charge of +1, and neutrons of mass 1 and charge 0.	
I know that isotopes are atoms of elements that have the same atomic number but different mass numbers.	
I can deduce the formula of an ionic compound from the formula of the positive and negative ions.	
I know that solid sodium chloride is a giant ionic lattice of positive ions strongly attracted to negative ions.	
I know that non-metals combine together by sharing electron pairs and that this is called covalent bonding.	
I can recognise the period to which an element belongs by the number of occupied electron shells in its atom.	
I can explain that group 1 elements have similar properties as they all have one electron in their outer shell.	
I can describe how to use a flame test to identify the presence of lithium, sodium and potassium compounds.	
I can identify the metal halide formed when a group 1 element reacts with a group 7 element.	
I know that when chlorine reacts with solutions of metal halides, it displaces bromides as Br_2 and iodides as I_2.	
I know that carbonates of transition metals decompose on heating to make the metal oxide and carbon dioxide.	
I know how using sodium hydroxide solution identifies Cu^{2+}, Fe^{2+} and Fe^{3+} as different coloured precipitates.	
I know that metals have high melting points and boiling points due to strong metallic bonds.	
I know potential benefits of superconductors include super-fast electronic circuits and powerful electromagnets.	
I know that the water purification process includes filtration, sedimentation and chlorination to kill microbes.	
I know that the reaction of barium chloride with sulfates is an example of a precipitation reaction.	
I am working at grades D–C	

C3 and C4 Grade booster checklist

P3	
I can calculate average speed and describe how average speed cameras work.	
I can interpret the shape of distance–time graphs.	
I can recognise situations that involve acceleration.	
I can describe the trends shown by speed–time graphs.	
I can recognise when forces affect speed and can calculate force from knowledge of mass and acceleration.	
I can describe thinking distance, braking distance and overall stopping distance.	
I can describe how energy is transferred when work is done and calculate the amount of work done.	
I can describe power as being the rate at which work is done.	
I can understand that kinetic energy depends on both mass and speed.	
I can describe the different fuels used for transport and interpret basic data on fuel consumption.	
I can calculate momentum and recall that sudden changes of momentum lead to forces which cause injury.	
I can describe the roles played by different safety features of modern cars.	
I can recognise how frictional forces affect movement.	
I can explain how falling objects reach a terminal speed.	
I can recognise that objects have GPE because of their mass and position in the Earth's gravitational field.	
I can recognise everyday examples in which objects use GPE.	
I am working at grades G–E	

I can interpret the relationship between initial speed, final speed, distance and time.	
I can interpret the gradient of a speed–time graph.	
I can draw and interpret the shapes of speed–time graphs.	
I can describe the significance of positive and negative acceleration and calculate acceleration.	
I can describe and interpret the relationship between force, mass and acceleration.	
I can explain the factors that affect thinking distance, braking distance and their implications on road safety.	
I can calculate weight from knowledge of mass and gravitational field strength.	
I can calculate power knowing the time it takes to perform work.	
I can use and apply the equation $KE = \frac{1}{2}mv^2$.	
I can explain how electrically powered cars do cause pollution but not at their point of use.	
I can describe the relationships between momentum, mass, velocity, force and time.	
I can explain how and why crumple zones, seat belts and air bags reduce injuries.	
I can explain the motion of a falling object in terms of balanced and unbalanced forces.	
I can recognise that all objects fall with the same acceleration at a point on the Earth's surface.	
I can use the equation GPE = mgh and interpret examples of energy transfer from GPE to KE.	
I can interpret the energy transfers of a roller coaster and the effects of mass and speed on kinetic energy.	
I am working at grades D–C	

P4	
I can describe how insulating materials can become charged and state that there are two kinds of charge, positive and negative.	
I can describe how you get an electric shock if you are charged and become earthed.	
I can recall that electrostatics is useful for dust precipitators, defibrillators and crop and paint spraying.	
I can describe how resistors are used to change the current in a circuit.	
I can identify the colours of the live, neutral and earth wires in a three-pin plug.	
I can describe why fuses and circuit breakers are used.	
I can recall that ultrasound is a longitudinal wave.	
I can recognise the features of a longitudinal wave including the wavelength, a compression and a rarefaction.	
I can recall the main sources of background radiation.	
I can compare the properties of x-rays and gamma rays.	
I can recall that medical radioisotopes are produced by placing materials in a nuclear reactor.	
I can describe uses of nuclear radiation in medicine and recall that only beta and gamma rays can pass through skin.	
I can give examples of the use of radioisotopes as tracers, in smoke detectors and for dating rocks.	
I can describe the role of a radiographer.	
I can describe the main stages in the production of electricity and recall that nuclear fission produces radioactive waste.	
I can describe the difference between fission and fusion and recall the 'cold fusion' controversy.	
I am working at grades G–E	

I can state that like charges repel and opposite charges attract.	
I can describe electrostatic phenomena in terms of transfer of electrons which have a negative charge.	
I can describe problems and dangers caused by static electrical charge.	
I can explain how static electrical charge can be useful in dust precipitators, defibrillators, crop spraying and paint spraying.	
I can calculate resistance and describe how the resistance of a wire can be changed.	
I can describe the functions of the three wires and the fuse in a three-pin plug.	
I can describe the features of a longitudinal wave.	
I can recall what ultrasound is and describe how it is used to scan the body and to break down kidney and other stones.	
I can explain and use the concept of half-life of radioactive isotopes.	
I can describe radioactivity as naturally occurring radiation from the nucleus of an unstable atom and give examples of sources of background radiation.	
I can recall the natures of alpha, beta and gamma radiation.	
I can describe how radioactive tracers are used in industry and in hospitals.	
I can describe other uses of radioactive isotopes such as in smoke detectors or radioactive dating.	
I can describe how x-ray images are produced.	
I can describe how nuclear fission is used in power stations and how the reaction is controlled.	
I can describe nuclear fusion and explain why it is difficult to produce power from it.	
I am working at grades D–C	

Modern periodic table

Group 1	Group 2												Group 3	Group 4	Group 5	Group 6	Group 7	Group 0
							1 **H** hydrogen											4 2 **He** helium
7 3 **Li** lithium	9 4 **Be** beryllium												11 5 **B** boron	12 6 **C** carbon	14 7 **N** nitrogen	16 8 **O** oxygen	19 9 **F** fluorine	20 10 **Ne** neon
23 11 **Na** sodium	24 12 **Mg** magnesium												27 13 **Al** aluminium	28 14 **Si** silicon	31 15 **P** phosphorus	32 16 **S** sulfur	35 17 **Cl** chlorine	40 18 **Ar** argon
39 19 **K** potassium	40 20 **Ca** calcium	45 21 **Sc** scandium	48 22 **Ti** titanium	51 23 **V** vanadium	52 24 **Cr** chromium	55 25 **Mn** manganese	56 26 **Fe** iron	59 27 **Co** cobalt	59 28 **Ni** nickel	64 29 **Cu** copper	65 30 **Zn** zinc		70 31 **Ga** gallium	73 32 **Ge** germanium	75 33 **As** arsenic	79 34 **Se** selenium	80 35 **Br** bromine	84 36 **Kr** krypton
85 37 **Rb** rubidium	88 38 **Sr** strontium	89 39 **Y** yttrium	91 40 **Zr** zirconium	93 41 **Nb** niobium	96 42 **Mo** molybdenum	99 43 **Tc** technetium	101 44 **Ru** ruthenium	103 45 **Rh** rhodium	106 46 **Pd** palladium	108 47 **Ag** silver	112 48 **Cd** cadmium		115 49 **In** indium	119 50 **Sn** tin	122 51 **Sb** antimony	128 52 **Te** tellurium	127 53 **I** iodine	131 54 **Xe** xenon
133 55 **Cs** caesium	137 56 **Ba** barium	139 57 **La** lanthanum	178 72 **Hf** hafnium	181 73 **Ta** tantalum	184 74 **W** tungsten	186 75 **Re** rhenium	190 76 **Os** osmium	192 77 **Ir** iridium	195 78 **Pt** platinum	197 79 **Au** gold	201 80 **Hg** mercury		204 81 **Tl** thallium	207 82 **Pb** lead	209 83 **Bi** bismuth	210 84 **Po** polonium	210 85 **At** astatine	222 86 **Rn** radon
223 87 **Fr** francium	226 88 **Ra** radium	227 89 **Ac** actinium																

Modern periodic table. You need to remember the symbols for the highlighted elements.

B1 Answers

B1 Understanding organisms

Page 134 Fitness and health

1 a i 300/200

    ii Greater/higher blood pressure needed; to pump blood up long neck

  b mmHg

  c Diastolic; systolic

(Any order)

2 Take more regular exercise; eat a healthier diet; reduce salt intake; reduce saturated fat intake; drink less alcohol; avoid stress; don't smoke; maintain a healthy weight

(Any two)

3 a Incidence of heart disease in women has reduced; incidence of heart disease in men has increased; incidence of strokes in both women and men has greatly increased

  b Use worldwide results; use longer time period; include other measurements of fitness/health

(Any two)

4 a Being fit is the ability to do exercise – this will not stop/kill pathogens

  b Strength; stamina; cardiovascular efficiency

(Any two)

Page 135 Human health and diet

1 a Meat to protein, sugar to carbohydrates, fruit juice to vitamins; carbohydrates to high energy source, protein to growth, vitamins to vitamin C, minerals to iron

  b Fats; fibre; water

(Any two)

  c i Very overweight

    ii Arthritis; heart disease; diabetes; breast cancer

(Any two)

2 a More fibre; less cholesterol; less fat

(Any two)

  b Less protein; less energy

3 a 40 000 g = 40 kg; EAR = 40 × 0.6 = 24 g

  b Low self-esteem; poor self-image; desire for 'perfection'; lack of knowledge

(Any 2)

4 a Carbohydrates

  b Amino acids

Page 136 Staying healthy

1 a Fungi to athlete's foot; bacteria to cholera; viruses to flu; protozoa to malaria

  b Not infectious; not caused by a pathogen; inherited/genetic cause

(Any two)

2 To see if they work; to see if they are safe

3 a Pathogens; toxins

  b Antigens; antibodies

4 a Carries a pathogen; so spreads disease

  b Parasite is mosquito or *Plasmodium;* host is human

5 a i At 20 days the level of passive immunity is more/twice that of active immunity

    ii after 60 days there is no immunity from passive immunity/active immunity still high

  b In passive immunity the body receives antibodies; this results in a high immunity level being quickly reached; in active immunity the body makes its own antibodies continually; so the level of antibodies and immunity remains high

Page 137 The nervous system

1 Tongue; chemicals in air; ears

2 a 1 is lens; 2 is retina; 3 is cornea

  b Reflected should be refracted; optic nerve should be retina; spinal cord should be brain

  c Narrower field of view; poorer judge of distance

3 An electric impulse; CNS

4 a B/axon

  b C

Page 138 Drugs and you

1 a Addicted; withdrawal symptoms

  b Tars; nicotine

  c Cigarette smoke stops cilia working; cilia are found on some epithelial cells; lining the trachea/bronchi/bronchioles; mucus traps dust/particles in air breathed in; mucus now not moved upwards by cilia; particles irritate, causing coughing

(Any four)

  d i 251 (the numbers of all the types added together)

    ii Fewer men smoke; people now know risks/warnings on cigarette packets/smoking ban in public areas

2 Depressant = alcohol; painkiller = paracetamol; stimulant = caffeine; hallucinogen = LSD

(All four for three marks)

3 Dangerous/can cause accidents; impairs judgement; impairs muscle control; can result in blurred vision; can result in drowsiness

(Any four)

Page 139 Staying in balance

1 a i 37 °C

    ii Clinical thermometer; sensitive strip; digital recording probe; thermal imaging

(Any 2)

  b Heat gained by respiring and less sweating; heat lost by sweating a lot, wearing few clothes

  c Sweat more; therefore lose too much water

  d Maintaining a constant internal environment

2 a Insulin; pancreas; blood; slower than

  b Type 2 diabetes caused by too little insulin or cells not reacting to it; amounts of carbohydrates eaten can be altered; to suit activity

  c i Glucose dissolves in the blood

    ii Diabetic person's pancreas not working/not making (enough) insulin; so glucose remains in blood; slowly removed by kidneys

(Any two)

B2 Answers

Page 140 Controlling plant growth

1 Get more light; make more glucose

2 Set up described: two sets of germinating cress seeds in container with a hole; Control: one set receiving light from all directions, the other from only one direction; Results: cress receiving light from all directions grew straight upwards, cress with light from only one direction grew curved towards the light; Conclusion: plant stems grow towards the light/stems are positively phototropic

3 a Makes roots grow; because rooting powder contains auxins/plant hormones

  b i Only kills/affects growth of broad-leaved plants; does not affect narrow-leaved plants such as grasses

   ii Contains plant hormones; which makes plant/roots grow too fast

4 Positively; gravity; geotropic; auxins

Page 141 Variation and inheritance

1 Body mass; height; intelligence
(If more than 2 boxes ticked, deduct 1 mark for each incorrect answer)

2 Matching pairs; sickle cell anaemia

3 Nucleus; genes; genetic

4 a Sudden change in gene/chromosome

  b During gamete formation; genes mixed up; fertilisation; recombination of genes from two parents
(Any two)

5 23 chromosomes in egg; 23 chromosomes in sperm; 46 chromosomes in Charlotte's cell

Page 142 B1 Extended response question

5–6 marks
The answer includes an explanation of both how and why the blue whale controls its body temperature. The similarity of whale and human temperature control is realised and temperature of about 37 °C is stated. The answer includes information on heat gain/retention (less blood flow near skin surface, respiration, blubber for insulation) as well as heat loss (more blood flow). Importance of balancing internal temperature (homeostasis) is realised by links to enzyme action and heat stroke and hypothermia. The answer links up migration to hot and cold conditions (oceans). Changes in the blood flow near the skin are described, as well as their corresponding effects (restricted blood flow means less heat loss). All information in answer is relevant, clear, organised and presented in a structured and coherent format. Specialist terms are used appropriately. Few, if any, errors in grammar, punctuation and spelling.

3–4 marks
The answer refers to changes in blood vessels in differing conditions but effects not clearly explained. Some references to heat loss and heat gain. References to actual body temperature and links to enzyme action are vague or missing. Answer not well targeted at the whale, i.e. no references to blubber or differing sea temperatures. For the most part the information is relevant and presented in a structured and coherent format. Specialist terms are used for the most part appropriately. There are occasional errors in grammar, punctuation and spelling.

1–2 marks
The answer simply describes changes in the blood vessels. References to heat gain/loss confused. Answer may be simplistic. There may be limited use of specialist terms. Errors of grammar, punctuation and spelling prevent communication of the science.

0 marks
Insufficient or irrelevant science. Answer not worthy of credit.

B2 Understanding our environment

Page 143 Classification

1 a Multicellular/feed on other organisms

  b Phylum; class; order; genus; species

2 a Similar; habitat; variation

  b A group of organisms that can interbreed; to produce fertile offspring

  c i Bobcat and ocelot
   ii Both belong to the same genus (*Felix*)

3 One type of organism gradually evolved into another; therefore there are always going to be intermediate organisms

Page 144 Energy flow

1 a i A feeding level in a food chain/a group of organisms that eat the same type of food
   ii Algae/bacteria
   iii May go down: fewer mice for owls to eat, so they eat more hedgehogs; May go up: fewer beetles eaten by mice, so more food for hedgehogs

  b i The dry mass of living material; at each trophic level
   ii Grass can feed many beetles; pyramid of numbers does not take into account the size of the organism

2 a The Sun/sunlight

  b Respiration

  c i Keep the cows in warm conditions; restrict their movement
   ii Have more energy available for growth

Page 145 Recycling

1 Decay; nitrogen; recycled

2 a Combustion; respiration; photosynthesis

  b Less oxygen in waterlogged soils; decomposers cannot respire

3 a 78%

  b It is unreactive

  c To make proteins

Page 146 Interdependence

1 a In some areas, the reds and the greys have lived together for many years

  b Food/acorns

  c Mates

2 a It shows the numbers of predator organisms; and the numbers of prey

b The numbers go up when there are fewer owls because fewer lemmings are eaten; when the owl numbers increase the lemming numbers drop as more are eaten

3 a Oxpeckers get food by eating the insects; buffalo get their parasites removed

b Mutualism

4 a Mites

b Bees gain nectar; flowers get pollinated

Page 147 Adaptations

1 a Eyes on the front of their head

b i Can make them both look forward; judge distances

ii Can look all around; spot snakes trying to approach

iii Green coloured like leaves; camouflaged

2 a Thick fur for insulation; layer of blubber under the skin; wide feet stop it sinking into the snow; claws to hold on to the ice; white colour for camouflage; small ears lose less heat; large body to retain heat
(Any three)

b Hibernation; body reactions slow down; conserve food reserves; less food is available in winter
(Any three)

3 a Less water loss through leaves; protect cacti from being eaten

b Sun basking/absorb Sun's heat; to warm up body

Page 148 Natural selection

1 a The plants with more spines will survive; less likely to be eaten

b Evolution

c Become extinct

2 a Can produce an enzyme to digest the acorns

b Some of the red squirrels would have a mutation; this would allow them to digest acorns; they are more likely to survive; pass on this gene; over many generations the population will all be able to digest
(Any four)

3 Some bacteria are resistant to antibiotics; they can survive and reproduce

4 Worried that people would disagree with him; most people believed that God created all organisms as they are now

Page 149 Population and pollution

1 a i Burning fossil fuels/forests

ii More demand for energy/population is increasing

iii Increase in temperatures/global warming; rising sea levels/flooding; drought
(Any two)

b i CFCs

ii UV light causes skin cancer

2 a Finite

b Minerals; fossil fuels

c Exponential

3 a They can be killed by pollution

b Indicator species

c Mussels, damselfly larvae and bloodworms all survive in polluted water; if it was clean you would also find mayfly and stonefly larvae

Page 150 Sustainability

1 a i Extinct

ii Habitat destroyed; overhunted; competitor introduced; disease; pollution
(Any two)

b i Captive breeding; protect habitat; ban hunting; education; artificial ecosystems
(Any two)

ii More tourism; better transport; other resources
(Any two)

2 a For: provides food/jobs; Against: may lead to their extinction

b Areas too large to police

3 a A resource that can be produced at the same rate that it is being used/can be taken out of the environment without running out

b Restricting numbers stops too many being killed; preventing killing small fish allows them to mature to breed

Page 150 B2 Extended response question

5–6 marks

The answer includes a description of how energy from sunlight is converted to chemical energy and passes through the food chain by eating. It also includes a description of how energy is lost from a food chain and some realisation that different amounts of energy are lost between different trophic levels. Answers are backed up by quoting of figures. All information in answer is relevant, clear, organised and presented in a structured and coherent format. Specialist terms are used appropriately. Few, if any, errors in grammar, punctuation and spelling.

3–4 marks

The answer refers to photosynthesis and includes at least one example of how energy is lost from food chains. At least one figure is quoted. For the most part the information is relevant and presented in a structured and coherent format. Specialist terms are used for the most part appropriately. There are occasional errors in grammar, punctuation and spelling.

1–2 marks

An incomplete description, either showing an appreciation of how energy passes or how it may be lost but not both. Answer may be simplistic. There may be limited use of specialist terms. Errors of grammar, punctuation and spelling prevent communication of the science.

0 marks

Insufficient or irrelevant science. Answer not worthy of credit.

C1 Carbon chemistry

Page 152 Making crude oil useful

1 a Coal, crude oil, (natural) gas

b They take a long time to make and they are used up faster than they are formed

2 a Each fraction has a different boiling point/range

  b Propane and butane

  c Petrol; diesel; paraffin; heating oil; fuel oils; bitumen
    (2 marks – 4 correct, 1 mark – 2 or 3 correct)

  d

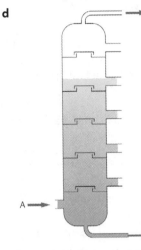

    i A At the bottom of the tower (left-hand side)

    ii B It 'exits' through the bottom of the tower

    iii C At the top of the tower

    iv The fraction with a low boiling point, LPG

3 Oil slicks can damage birds' feathers causing death; the use of detergents to clean it up causes problems for wildlife

4 a Liquid paraffin (alkane) on mineral fibre, aluminium oxide, very strong heat, water, hydrocarbon gas (alkene)

  b Catalyst; high temperature

  c An alkene has a double covalent bond between carbon atoms, an alkane has a single covalent bond

  d Polymers

Page 153 Using carbon fuels

1 a Cost

  b Coal is bulky and dirty whereas petrol is volatile; petrol produces less acid fumes than coal

2 a To release useful heat energy; oxygen

  b Carbon dioxide; water

  c **i** Hydrocarbon fuel + oxygen → carbon dioxide + water
      (1 mark for reactants, 1 mark for products)

    ii The candle is burnt; and the products are passed through test reagents (if they are positive, then carbon dioxide and water are formed)

3 a **i** Shortage of oxygen

    ii Carbon monoxide/carbon/soot; water
      (1 mark for carbon monoxide or carbon/soot,
      1 mark for water)

  b Less soot is made; more heat is released; toxic carbon monoxide gas is not produced

Page 154 Clean air

1 a Water vapour

  b Nitrogen; oxygen; carbon dioxide

  c 78%; 21%; 0.035%

2 a

	Photosynthesis	Respiration	Combustion
carbon dioxide	goes down	goes up	goes up
oxygen	goes up	goes down	goes down

(1 mark for each correct column)

  b Plants photosynthesise using up carbon dioxide; <u>photosynthesis</u> produces oxygen; animals respire using up oxygen; animal <u>respiration</u> produces carbon dioxide; fuels undergo combustion using up oxygen; fuel <u>combustion</u> produces carbon dioxide
    (Answer must include underlined terms)

3 a Kills plants; kills aquatic life (fish); erodes stonework; corrodes metals   *(Any 3)*

  b When there is incomplete combustion in petrol-powered cars, it produces poisonous gases such as carbon monoxide; oxides of nitrogen can contribute to photochemical smog and acid rain; if there is sulfur in the fuel, acid rain can be made

Page 155 Making polymers

1 a Hydrogen and carbon

  b A, B, C

  c Hydrocarbon (alkane)

  d It contains an oxygen atom

  e It contains a double bond

  f The bromine water is orange; when the alkene is present, the water is decolourised/it turns colourless

2 a When many small monomers join together; to make a large molecule called a polymer

  b Poly(propene)

  c Styrene

  d C

  e High pressure; catalyst

Page 156 Designer polymers

1 a It is waterproof and can be made into a fibre

  b It is waterproof, rigid and a heat insulator

  c Bags need to be flexible and waterproof which are properties of both nylon and poly(ethene)

  d Any reasonable example of use; and suitability. For example:

Use	Suitability
Contact lens	flexible
Drain pipe	rigid
Wound dressing	waterproof

  e It keeps in water vapour from body sweat

2 a They do not break down by bacterial action

  b **i** Wastes valuable land

    ii Creates toxic gases

    iii Difficult to sort different polymers

  c They do not have to be disposed of in landfill sites or burnt; they can decay by bacterial action or be dissolved

Page 157 Cooking and food additives

1 a Cooking is a chemical change as it is <u>irreversible</u> and makes uncooked food into a new <u>substance</u>.

There is an <u>energy</u> change and the process cannot be <u>reversed</u>.

(1 mark for each two underlined words correctly explained)

b Change shape (permanently); the process is called denaturing

2
antioxidants — prevent reaction with oxygen
food colours — improve the appearance
flavour enhancers — improve the taste
emulsifiers — helps mix oil and water

(2 marks – 3 correct, 1 mark – 2 correct)

3 a It decomposes on heating to give off carbon dioxide

b Carbon dioxide; limewater turns cloudy/milky

c Sodium hydrogencarbonate → sodium carbonate + carbon dioxide + water

d $2NaHCO_3 \rightarrow Na_2CO_3 + CO_2 + H_2O$
(1 mark for all 4 symbols, 1 mark for balancing the equation)

Page 158 Smells

1 a Lavender; rose oil; any reasonable named flower oil; musk

(Any 1)

b Synthetic

c i Acid + alcohol → ester + water

ii Put some alcohol; in a test tube; add an equal amount of ethanoic acid; heat in a boiling water bath; diagram of test tube in a heated beaker of water (accept also flask and condenser as a reflux apparatus)

2
evaporate easily — its particles can reach the nose
non-toxic — it does not poison people
insoluble in water — it cannot be washed off easily
not irritate the skin — it can be put directly on the skin
does not react with water — it does not react with perspiration

(4 marks – 5 correct, 3 marks – 4 correct, 2 marks – 3 correct, 1 mark – 2 correct)

3 Water cannot be used to remove varnish from nails, water is not a <u>solvent</u> for nail varnish. Nail varnish is <u>insoluble</u> in water. Nail varnish is <u>soluble</u> in nail varnish remover. A solvent dissolves a <u>solute</u> to make a <u>solution</u>.
(3 marks – 5 correct, 2 marks – 4 correct, 1 mark – 3 correct)

4 a Solution

b Perfumes and solvents

5 They must be safe in immediate use; they must show no long-lasting or future damage to health

Page 159 Paints and pigments

1 a
solvent — thins the paint making it easier to spread
binding medium — sticks the pigment in the paint to the surface
pigment — substance that gives the paint its colour

(2 marks – 3 correct, 1 mark – 2 correct)

b A paint where the pigment powder is dispersed (spread) through the oil (a solvent is often added that dissolves oil)

c i Protection/attractiveness

ii Attractiveness

iii Protection and attractiveness

iv Interesting images/representation/impressions/any suitable answer

d Particles/solute are dispersed through a liquid/ solvent

e The solvent evaporates and the paint dries to form a continuous film

2 a Change colour when heated or cooled

b Cups/kettles/child's cutlery/child's bath toy/any reasonable use

c Answers could include: a cup, so that it can change colour if it is too hot to touch; a baby's bath toy so that it can provide a warning if the water is too hot

3 They absorb energy; the energy is released as light over a long period of time

Page 160 C1 Extended response question

5–6 marks
Comprehensive descriptions of the experiment including: acid and alcohol used to make an ester; the acid and alcohol need to be heated for some time so a flask and condenser is used (reflux). At least three properties of perfumes should be listed from the following list: evaporates easily, non-toxic so it does not poison, does not react with water so it does not react with perspiration, does not irritate the skin so it can be used on skin safely.
An explanation of the removal of nail varnish using the word 'soluble' or 'insoluble' (nail varnish is insoluble in water, but soluble in nail varnish remover; nail varnish remover is a solvent).
All information in the answer should be relevant, clear, organised and presented in a structured and coherent format. Specialist terms should be used appropriately. There should be few, if any, errors in grammar, punctuation and spelling.

3–4 marks
One point made in the description of the experiment. At least two properties of perfumes. Some attempt at the explanation of the removal of nail varnish.
The information should generally be relevant and presented in a structured and coherent format. Specialist terms should mostly be used appropriately. There may be occasional errors in grammar, punctuation and spelling.

1–2 marks
One point made in the description of the experiment. Two properties of perfumes listed or one property of perfume and an attempt at the explanation of the removal of nail varnish.
Answers may be simplistic. There will be limited use of specialist terms. Errors of grammar, punctuation and spelling are likely to prevent effective communication of the science.

0 marks
Insufficient or irrelevant science such as repeating the question. Answer not worthy of credit.

C2 Chemical resources

Page 161 The structure of the Earth

1 a Thin, rocky crust; mantle; iron core

b Crust; top part of the mantle

c Crust too thick to drill through; need to rely on infrequent seismic waves or man-made explosions

2 a Plates move very slowly (2.5 cm per year); continents move over millions of years

b Volcanoes/volcanic activity; earthquakes

c The theory explains the evidence; it has been discussed/tested by many scientists

3 a Quick/rapid cooling; results in small crystals; slow cooling; makes bigger crystals

b Slowly with runny lava; violently with thick lava

c Because the ash is good for growing crops/the land is very fertile with many soil nutrients

d To predict eruptions/to save lives; to find out information about the Earth's structure

Page 162 Construction materials

1 a Limestone; marble; granite

b Landscape is destroyed and has to be reconstructed; increased noise, dust and traffic

c Limestone, marble, granite

d

Building material	aluminium	brick	glass
Raw material	bauxite (aluminium) ore	clay	sand

e Limestone and clay are heated together

2 a Calcium carbonate

b Calcium oxide; and carbon dioxide

c Concrete is made by mixing cement, sand and small stones/aggregate with water and allowing the mixture to set; it can be strengthened by adding steel bars/grids

d Calcium carbonate → calcium oxide + carbon dioxide

Page 163 Metals and alloys

1 a Heated; with carbon

b Removal of oxygen from a substance

c Electrolysis

d Label to left (negative) side

e Advantages: the cost to melt it is low; people earn money from selling it; less mining needed, so there is less noise pollution and dust, and fewer trucks; the cost of copper is kept down
(Any 2)

Disadvantages: it leads to fewer jobs in mining; some items are hard to separate; pure and less pure copper must not be mixed which can be difficult to check; some items contain hardly any copper so it is not worthwhile; the separation process can create pollution; it is hard to persuade people to recycle
(Any 2)

2 a Good electrical conductivity; low density; high tensile strength

b i musical instrument — amalgam
joins electrical wires — brass
tooth fillings — solder

ii amalgam — contains copper and zinc
brass — contains mercury
solder — contains lead and tin
(2 marks – 3 correct, 1 mark – 2 correct)

Page 164 Making cars

1 a Iron, oxygen, water

b It happens with the addition of oxygen

c Salt water accelerates rusting in steel and so car bodies made from steel rust more quickly when salt is put on the roads

d Aluminium forms a protective oxide layer

e Iron + oxygen + water → hydrated iron(III) oxide
(1 mark for reactants, 1 mark for products)

2 a Steel is harder and is less likely to corrode

b Lighter car than same size steel car; will corrode less

c

Material and its use	Property
copper in electrical wires	electrical conductivity
plastic bumpers	cleanliness/density/probable corrosion
PVC wire covering	electrical conductivity/ flexibility

d Transparent; strong; does not shatter
(Any 2 for one mark)

3 a Aluminium

b Moist acidic air

c Aluminium; the other materials are affected more by the atmosphere according to the table

Page 165 Manufacturing chemicals – making ammonia

1 a Nitrogen comes from the air; hydrogen comes from natural gas or cracking of oil fractions

b $\rightleftharpoons$

c High pressure; 450 °C temperature; (iron) catalyst; recycling unreacted starting materials
(Any 3)

d $N_2 + 3H_2 \rightleftharpoons 2NH_3$
(1 mark for all 3 symbols, 1 mark for balancing the equation)

2 a 50%

b The higher the pressure, the higher the yield

3 a Energy; starting materials; wages; equipment/plant/ automation; catalysts
(Any 3)

b Higher pressure and temperature means higher energy cost/other links with factors above

c Ammonia is used to make nitrogen fertilisers; fertilisers are needed to increase crop yield; as world population increases more fertiliser is needed; the amount of land available for growth decreases, so the existing land needs to produce more crops
(Any 3)

Page 166 Acids and bases

1 a Less than 7

b A soluble base (accept a solution of pH above 7)

c i Red   **ii** Blue

2 a Neutralisation

b Acid, water

c Hydrogen ion/H^+ ion

d The greater the concentration of H^+ ions, the lower the pH

3 a The pH at the start is low, as alkali is added to the acid the pH rises

b The colour starts as purple showing it is alkaline; as the acid neutralises the alkali the pH falls and the colour changes to blue; when neutral, the pH = 7 and the colour is green

4 a Copper carbonate + sulfuric acid → copper sulfate + water + carbon dioxide

(1 mark for reactants, 1 mark for products)

b i Calcium sulfate

ii Potassium nitrate

iii Sodium chloride

iv Copper nitrate

Page 167 Fertilisers and crop yields

1 a Improve crop yield

b Through the roots

c Nitrogen; phosphorus; potassium

d Potassium; nitrogen

e So substances are small enough to get into the roots

2 a They replace/provide essential elements/nutrients and increase the food supply (accept, they provide nitrogen needed to make plant proteins resulting in increased growth)

b If too much fertiliser is used it can be washed off fields sometimes causing death to aquatic life (the process is called eutrophication)

3 a Burette; measuring cylinder; conical flask

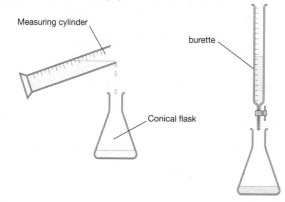

Measuring cylinder

burette

Conical flask

b Ammonium nitrate; ammonium phosphate; ammonium sulfate; urea

(Any two for 1 mark)

c i Ammonium nitrate

ii Potassium phosphate

Page 168 Chemicals from the sea – the chemistry of sodium chloride

1 a Seawater; salt deposits (accept 'salt pans')

b Mining rock salt from the ground; solution mining

c Subsidence of houses; brine can escape and affect habitats; noise; dust; trucks

(Any 2)

2 a Hydrogen; chlorine

b Chlorine bleaches damp litmus paper

c i Chlorine

ii Hydrogen

iii Sodium hydroxide

d To prevent unwanted reactions/so the electrodes do not dissolve

3 a Flavouring; preservative

b Sterilising water supplies/swimming pools; making household bleach; manufacturing polymers/plastics such as PVC; making solvents

(Any 2)

c Used in margarine production

d To make soap

e Sodium hydroxide; chlorine

Page 169 C2 Extended response question

5–6 marks

The measuring cylinder, funnel and flask should all be labelled. Any three of potassium nitrate, potassium sulfate, ammonium nitrate or ammonium sulfate should be given.

All information in the answer should be relevant, clear, organised and presented in a structured and coherent format. Specialist terms should be used appropriately. There should be few, if any, errors in grammar, punctuation and spelling.

3–4 marks

The apparatus should be labelled and at least one fertiliser name predicted.

The information should generally be relevant and presented in a structured and coherent format. Specialist terms should mostly be used appropriately. There may be occasional errors in grammar, punctuation and spelling.

1–2 marks

At least two pieces of apparatus should be labelled, and one name of a fertiliser predicted.

Answers may be simplistic. There will be limited use of specialist terms. Errors of grammar, punctuation and spelling are likely to prevent effective communication of the science.

0 marks

Insufficient or irrelevant science. Answer not worthy of credit.

P1 Energy for the home

Page 170 Heating houses

1 a Energy; joules

b Energy flows from warm to cooler body; temperature of warmer body falls

2 a 8 minutes

  b i Specific heat capacity

    ii Energy needed = mass × specific heat capacity ×
temperature change = 0.5 × 3900 × 70; = 136 500 J

    iii To heat the beaker

3 a Melting ice/boiling water/any change of state

  b Specific latent heat

Page 171 Keeping homes warm

1 Trapped; insulator

2 a Particles in solid close together/particles in gas far
apart/no particles in vacuum; gap between glass
filled with gas or vacuum/ more difficult to transfer
energy than in a solid

  b i Air in foam is good insulator/reduces energy
transfer by conduction; air is trapped/unable to
move; reduces energy transfer by convection

    ii Energy from room is deflected back in winter;
energy from the Sun is reflected back in summer

  c Loft insulation; draught strip; curtains; carpets;
underlay

                                        (Any 1)

3 a Only 32% of energy input is useful as energy output

  b 0.32 × 9.5 = £3.04

                                        (2 marks)

  c Energy is lost up the chimney

Page 172 A spectrum of waves

1 a Radio; microwave; infrared; ultraviolet; x-rays;
gamma rays

                                        (Any 3)

  b Vertical arrow

2 a A on vertical peak to axis arrow

  b W on horizontal trough to trough arrow

  c Number of complete waves passing a point
each second

3 Reflections from both mirrors; reflected ray returns
along incident path

Page 173 Light and lasers

1 Semaphore; smoke signals; runner; any suitable

                                        (Any 1)

2 Dots and dashes – series of on/off signals; not
continuously variable signal

3 a More; water/glass/plastic/Perspex; less; air; incidence

  b i x angle of refraction greater than angle of
incidence; y ray glancing along water/air
boundary; z ray reflected back into water; angle
of reflection = angle of reflection by eye

    ii c marked as angle between ray and normal in
diagram y

4 Communication; dental treatment; weapon guidance;
surgery; light show

                                        (Any 2)

Page 174 Cooking and communicating using waves

1 a Infrared; absorb; spectrum; water

  b Microwave radiation is more penetrating than
infrared

2 a There is some evidence of heat energy being
transferred to the body; young people are more
likely to be affected

  b Microwaves need line of sight; there are no
obstructions in space

Page 175 Data transmission

1 a True; false; true

  b Digital

  c Any continuously varying waveform

2 Greater choice of programmes; interact with
programmes; information services

                                        (Any 1)

3 3–5 reflections along length of fibre; ray reflected from
surface with equal angles (by eye)

Page 176 Wireless signals

1 Less refraction at higher frequencies

2 Can be used anywhere/portable   *(Any 1)*

3 Reflected

4 The radio station is broadcasting on the same
frequency/the radio waves travel further because of the
weather conditions

Page 177 Stable earth

1 a Seismometer

  b Transverse – S and longitudinal – P; travels through
solid – P and S; travels through liquid – P

2 a Ultraviolet radiation; cells in skin produce melanin

  b 20 × 15; 300 minutes/5 hours

  c Repeat their readings/ask other scientists to repeat
the experiment

Page 178 P1 Extended response question

5–6 marks

A detailed description of how microwave radiation is used
to transmit a signal over a large distance to include line of
sight transmission, satellite receiving, amplifying and
retransmitting the signal. A discussion of possible dangers
from use of mobile phone near the head and that
children are more at risk because their bodies are still
developing.

All information in answer is relevant, clear, organised and
presented in a structured and coherent format. Specialist
terms are used appropriately. Few, if any, errors in
grammar, punctuation and spelling.

3–4 marks

A limited description of some of the details of how
microwave signals are transmitted over large distances.

Answers

A mention of possible dangers from microwaves and children more at risk.

For the most part the information is relevant and presented in a structured and coherent format. Specialist terms are used for the most part appropriately. There are occasional errors in grammar, punctuation and spelling.

1–2 marks

An incomplete description, stating that microwave radiation is used. A mention of possible dangers from microwaves.

Answer may be simplistic. There may be limited use of specialist terms. Errors of grammar, punctuation and spelling prevent communication of the science.

0 marks

Insufficient or irrelevant science. Answer not worthy of credit.

P2 Living for the future (energy resources)

Page 179 Collecting energy from the Sun

1 a Light; electricity

  b Photocells do not need fuel; do not need cables; need little maintenance; use a renewable energy source; cause no pollution or global warming

(Any 4)

2 a i Black absorbs heat

    ii Convection

  b i During the night the heated walls and floor radiate heat back into the room

    ii To benefit from the Sun's heat for most of the day

3 a Wind is caused by convection currents

  b Advantages – renewable; no pollution or global warming

(Any 2)

    Disadvantages – unreliable; noisy; spoil the landscape

(Any 2)

Page 180 Generating electricity

1 a Needle moves to the left

  b Use a stronger magnet; increase the number of turns in the coil; spin the coil faster

(Any 2)

  c In a power station the turbine spins the generator, electromagnets provide field, electromagnets turn inside coil where current is induced

2 a Generator produces electricity

  b In the form of heat

  c i 25%

    ii Heat losses in the boiler, generator, coils and cooling tower

Page 181 Global warming

1 a Methane; water vapour

  b Water vapour

2 a Increased it

  b Natural forest fires; volcanic eruptions; decay of dead plant and animal matter; evaporation from the oceans; respiration

(Any 4)

  c Burning fossil fuels; waste incineration; deforestation; cement manufacture

(Any 3)

3 a The rise in temperature of the Earth's atmosphere

(2 marks)

  b Increase – the smoke from factories reflects radiation from the town back to Earth

    Decrease – ash clouds from volcanoes reflect radiation from the Sun back into space

4 a Monitoring the atmosphere; sharing their findings with each other

  b They disagree on how much humans are contributing to global warming

  c On the basis of scientific evidence

  d Melting polar ice caps; rising sea levels

Page 182 Fuels for power

1 a 24 W   *(4 marks)*

  b 15p   *(4 marks)*

2 a Fossil fuels

  b i Wood; straw; manure

    ii Uranium; plutonium

  c Availability; risks associated with its production; its effects on the environment

3 a Reduces energy losses; reduced distribution costs; cheaper electricity for consumers

(Any 2)

  b Change size of AC voltage

Page 183 Nuclear radiations

1 True; true; true; false

2 a Use tongs; hold at arm's length; use for as little time as possible; return to storage box immediately after use

(Any 3)

  b Mutation of DNA in cells; cancer

3 a Radioactive tracer; radiotherapy

(Any 1)

  b Penetrates the body easily; can kill mutated cells

(Any 1)

  c The radiation ionises the oxygen and nitrogen atoms in air; this causes a very small electric current that is detected; when smoke fills the detector in the alarm during a fire, the air is not so ionised; the current is less and the alarm sounds

(Any 3)

4 Advantage – does not produce smoke; CO_2; contribute to global warming

(Any 1)

Disadvantage – produces radioactive waste; expensive to build and maintain

(Any 1)

5 a Waste can remain radioactive for thousands of years

  b Waste can be buried in landfill sites

Page 184 Exploring our Solar System

1
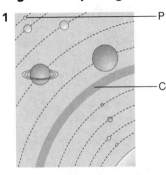
P

C

(2 marks)

2 a i A star is a ball of hot glowing gas giving out light
   ii A planet is a spherical object orbiting a star, which has cleared the neighbourhood of its orbit
   iii A meteor is made from grains of dust that burn up as they pass through the Earth's atmosphere

   b i A galaxy is a collection of stars
   ii A black hole is formed when a large star dies; you cannot see a black hole because light cannot escape from it

(Any 1)

3 a The Hubble space telescope looks for planets where conditions to support life may be found

   b To prevent astronauts being the blinded by the Sun's glare

   c Food; water; oxygen

Page 185 Threats to Earth

1 a It got colder; there were lots of fires; tsunamis flooded large areas

   b Mars; Jupiter

   c Unusual metals have been found near craters; fossils are found below the metal layer but not above; fossil layers have been disturbed by tsunamis
(Any 2)

2 Less dense rocks began to orbit; they joined together to form our Moon

3 a C on elliptical orbit

   b When it is being heated by the Sun

   c Increases as it approaches the Sun

4 a Asteroids or comets on a path which will bring them on a collision course with the Earth

   b So that we could track their position accurately; and deal with any that came too near Earth

Page 186 The Big Bang

1 a A single very hot point/ there was no Universe

   b Protons; hydrogen; helium

   c The ones furthest away

2 a Copernicus

   b By observations using a telescope

3 a By clouds and dust pulled together by gravity

   b First becomes a red giant; gas shells, called planetary nebula, are thrown out; the core becomes a white dwarf; then cools to become a black dwarf

Page 187 P2 Extended response question

5–6 marks
A detailed description of how uranium is used to provide the heat energy to heat up the water in a boiler and turn it into steam which is then used to turn turbines. These in turn are connected to generators where a coil spinning in a magnetic field generates electricity. Answer should include the fact that nuclear power stations carry the risk of incidents where radiation could leak as well as the need for careful disposal of the radioactive waste which will remain radioactive for a long time.

All information in the answer is relevant, clear, organised and presented in a structured and coherent format. Specialist terms are used appropriately. Few, if any, errors in grammar, punctuation and spelling.

3–4 marks
A limited description of how the power station works. A mention of possible dangers from radioactive waste.

For the most part the information is relevant and presented in a structured and coherent format. Specialist terms are used for the most part appropriately. There are occasional errors in grammar, punctuation and spelling.

1–2 marks
An incomplete description of the process in the power station. A mention of possible dangers from radioactivity. Answer may be simplistic. There may be limited use of specialist terms. Errors of grammar, punctuation and spelling prevent communication of the science.

0 marks
Insufficient or irrelevant science. Answer not worthy of credit.

B3 Living and growing

Page 188 Molecules of life

1 a In the cytoplasm – Mitochondria – The site of respiration; In the nucleus – Chromosomes – Contain the genetic code

   b To provide lots of energy; for contraction

2 a Gene; protein

   b i Circle around any one of the squares on the diagram
   ii Double helix
   iii Proteins are made in the cytoplasm; DNA cannot leave the nucleus

3 a Watson and Crick

   b C always equals G; T always equals A

Page 189 Proteins and mutations

1 a Collagen – a structural protein, haemoglobin – a carrier protein, insulin – a hormone

   b Amino acids

2 a A protein molecule; that speeds up a chemical reaction

   b A hole or groove on the enzyme; that the substrate fits into

   c Enzymes are specific; it would be the wrong shape to fit the active site

   d i As the temperature increases the reaction is faster; after a certain temperature any increase will decrease the rate
   ii 41–42 °C

3 a Mutation

b Radiation; certain chemicals

c Sometimes they produce an advantage; this is more likely to be passed on

Page 190 Respiration

1 a Muscle contraction; control of body temperature; protein synthesis

(Any two)

b Oxygen

c $6O_2 \rightarrow 6CO_2 + 6H_2O$

d It is respiring faster; so it needs to take in more oxygen/remove more carbon dioxide

e i As the horse runs faster there is an increase in lactic acid levels; this is slow at first but increases more rapidly after about 8m/s

ii At slower speeds respiration is mainly aerobic; at high speeds there is more anaerobic respiration and so more lactic acid is made

iii About 9 m/s

2 Measure the rate of oxygen consumption/rate of carbon dioxide production

Page 191 Cell division

1 a Each organism only has one cell

b Allows organism to be larger; allows for cell differentiation; allows organism to be more complex

(Any two)

2 a Mitosis

b The new cells are diploid; before cells divide, DNA replication takes place

3 Meiosis; double; fertilisation

4 The acrosome is needed to digest the egg membrane; this allows the nucleus of the sperm to enter

Page 192 The circulatory system

1 a White blood cell – defends against disease; red blood cell – transports oxygen; platelet – helps clot blood

b Disc shaped: larger surface area to take up oxygen quicker; No nucleus: more room to carry more haemoglobin/oxygen

c Haemoglobin

2 a Arteries; veins; capillaries

b Arteries carry blood away from the heart; veins carry blood back to the heart; capillaries are the site of exchange from the blood

3 a The lungs

b

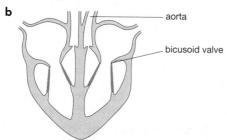

aorta

bicusoid valve

c The left ventricle has to pump the blood further; it has to generate more pressure

Page 193 Growth and development

1 a Cell wall; vacuole

b i It is smaller

ii A bacterial cell does not have a true nucleus; or mitochondria

2 a Adolescence/puberty

b Adolescence/puberty

c The girl is taller; because girls reach puberty/have their growth spurt at a younger age

3 a Cell differentiation

b Stem cells

4 a Plants grow throughout their life; plants only grow at specific parts of the plant

b The parts of a plant where cells divide

Page 194 New genes for old

1 a A and B

b Choose the parents that have closest to the desired characteristics and mate them together; choose the offspring that have closest to the desired characteristics and mate them; continue the process

c When two individuals that are closely related mate

2 a Genetic engineering/modification

b Many people are deficient in vitamin D; this can be made from beta-carotene; many of these people eat large amounts of rice

c The plants might be harmful to health in the long term; they might escape into the wild and affect food chains

3 a Cure people of the genetic disorder

b Gene therapy

Page 195 Cloning

1 a An identical genetic copy

b Identical twins

c Putting the nucleus from a body cell into an egg that has had its nucleus removed

d i The pigs have/contain human genes; so their organs are not rejected

ii Produce large quantities of pigs with desired characteristics such as meat yield/produce large numbers of pigs which produce human proteins

2 a B, A, E, D, C

b Advantages: produce large numbers quickly/if the parent produces good strawberries then the offspring will also; Disadvantage: all the strawberries will be identical and so all could be attacked by a disease

Page 196 B3 Extended response question

5–6 marks

The answer includes an appreciation that an inherited disorder is a mutation and it is caused by a change in a gene. Genes are passed on in sexual reproduction in the gametes. It also includes an explanation of the symptoms in terms of collagen's role as a structural protein. The particular susceptibility of arteries due to the high pressure is mentioned. All information in answer is relevant, clear, organised and presented in a structured and coherent format. Specialist terms are used appropriately. Few, if any, errors in grammar, punctuation and spelling.

3–4 marks
The answer refers to mutations and changes in DNA or genes but the inherited nature may not be described. The function of collagen as a structural protein is appreciated. For the most part the information is relevant and presented in a structured and coherent format. Specialist terms are used for the most part appropriately. There are occasional errors in grammar, punctuation and spelling.

1–2 marks
An incomplete description, either quoting the term mutations and the idea that a gene or DNA has been altered. Answer may be simplistic. There may be limited use of specialist terms. Errors of grammar, punctuation and spelling prevent communication of the science.

0 marks
Insufficient or irrelevant science. Answer not worthy of credit.

B4 It's a green world

Page 197 Ecology in the local environment

1 **a** Eating them; removing them; helping them by providing carbon dioxide; destroying them by polluting the area
(Any two)
 b i S. balanoides
 ii S. balanoides; A. modestus

2 Native woodland

3 **a** A pooter is a jar with two tubes; point one tube at a small animal; suck through other tube
 b i Capture-recapture
 ii $\frac{50 \times 60}{10}$; population size is 300
 iii Not all ladybirds are counted/population is sampled

Page 198 Photosynthesis

1 **a i** Carbon dioxide; water; oxygen
 ii Light/Sun
 b i Cell walls; starch; growth/repair; storage
 ii Energy source/respiration

2 van Helmont: plant growth could not be only due to soil minerals; Priestley: plants produce oxygen

3 **a** More light; higher temperature; more photosynthesis
 b Heater produces warmth; and carbon dioxide; shades removed so maximum light enters; watering system provides right amount of water
(Any three)

Page 199 Leaves and photosynthesis

1 Specialised for other functions; example given, e.g. epidermis for protection/xylem for water transport

2 Root hairs; stomata; stomata

3 **a** Large surface area; get more light; more photosynthesis; more food

 b Shorter distance for gases to diffuse/all cells can get light for photosynthesis; variety of pigments; vascular bundles/xylem and phloem

4 **a** A is upper epidermis; B is palisade; C is guard cell
 b i Palisade; mesophyll; xylem; phloem
(Any two)
 ii Eat cells containing chlorophyll; therefore parts will not be green

Page 200 Diffusion and osmosis

1 **a** Floppy/soft
 b Lost too much water/more water lost than taken up
 c Cell wall

2 Diffusion; carbon dioxide; more; enter

3 **a** Black circles move down and mix with white circles; white circles move up and mix with black circles
 b Diffusion
 c Random movement of particles; takes place from high to low concentrations

4 Water; partially-permeable; dilute; concentrated; random

Page 201 Transport in plants

1 **a** Root hairs; leaves; transpiration; balanced
 b Temperature; air movement; humidity
(Any two)

2 **a** 2, 9
 b Leaf A (fewer stomata open); no photosynthesis/stomata close to prevent too much water loss

3 **a** Water loss/transpiration
 b Leaf 4 has only lower surface exposed; it loses more weight/2.7 g instead of 1.0 g; it loses more water than Leaf 3, which has upper surface only exposed

Page 202 Plants need minerals

1 **a** Nitrogen; phosphorus; potassium
 b Magnesium
 c i N is 0.6 kg; P is 1.2 kg; K is 1.8 kg
 ii Yes: crop needs 0.25 kg; less than the 0.6 kg which is available
(All required for 2 marks)

2 **a** Nitrates; phosphates; potassium; magnesium
 b Cannot make proteins; for cell growth

Page 203 Decay

1 The chemicals in dead organisms are recycled

2 **a** Dead and decaying plant material
 b A temperature of 25 °C; plenty of oxygen
 c Earthworms; maggots; woodlice
 d For: avoids landfill/recycles minerals/saves buying fertilisers; Against, unsightly/smelly if not working properly (anaerobic respiration)/needs space or garden
 e Break up dead material; so increasing surface area for decay

3 a Canning; drying; adding salt/sugar; adding vinegar
(Any two)

  b i Temperature of 5 °C slows down microbial growth/reproduction; so slowing down decay

   ii Temperature of −22 °C kills most bacteria/microbes; so slowing down decay better than a refrigerator

Page 204 Farming

1 Insecticide to kills insects; fungicide to kills fungi; herbicide to kills weeds

2 a Sugar cane

  b Rice; pesticides kill weeds and insect pests (disease may not be caused by insects)

3 a Biological control

  b Insecticides can enter and accumulate in food chains; insecticides may harm other useful insects; some insecticides are persistent
(Any two)

  c Not native/from South America; no natural predators; lack of research on its diet/lack of trials
(Any two)

4 a Growing one of three or four crops in each of three or four fields; then growing another one in following years; so the cycle is complete in 3–4 years

  b The nitrogen fixed from air/soil; forms protein in plants; when dug back into the soil it breaks down into nitrates/fertilisers for the next crop

Page 205 B4 Extended response question

5–6 marks
The answer includes correct manipulation of data (soil lost 0.06 kg but plant had gained 75.24 kg).
An explanation for this, such as the increase in mass could not have come from the soil, is given. The candidate realises that the small loss in mass of the soil could be due to uptake of minerals.
The candidate names the process as photosynthesis. The next stage in understanding photosynthesis by Priestley is described (plants produce oxygen). A more modern description of photosynthesis is given using a word or symbol equation explaining that the gain in mass is due to the production of sugars using carbon dioxide from the air, water from the soil and light energy.
Some reference may be made about confidence in such data involving only one tree.
All information in answer is relevant, clear, organised and presented in a structured and coherent format. Specialist terms are used appropriately. Few, if any, errors in grammar, punctuation and spelling.

3–4 marks
The answer includes some data manipulation but it may not be correct. Some reference is made to the soil losing mass and the tree gaining mass.
The answer refers to photosynthesis but only briefly mentions Priestley's work and a present-day understanding. A word equation for photosynthesis may be included.
For the most part the information is relevant and presented in a structured and coherent format. Specialist terms are used for the most part appropriately. There are occasional errors in grammar, punctuation and spelling.

1–2 marks
The answer simply refers to the soil losing some mass and the plant gaining it.

Photosynthesis may be mentioned but a description of the process is not included.
Answer may be simplistic. There may be limited use of specialist terms. Errors of grammar, punctuation and spelling prevent communication of the science.

0 marks
Insufficient or irrelevant science. Answer not worthy of credit.

C3 Chemical economics

Page 206 Rate of reaction (1)

1 Rusting

2 a i Flask
   ii Gas syringe

  b i 28 cm³
   ii cm³/s
   iii It decreases/drops to zero
   iv Magnesium

3 a 15 seconds

  b Reaction 1

  c Any answer between 1 and 2 seconds

Page 207 Rate of reaction (2)

1 a 55 cm³

  b

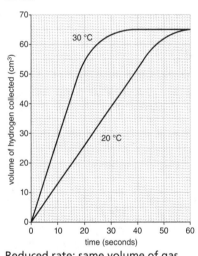

Reduced rate; same volume of gas

  c As the temperature decreases, the particles have less kinetic energy; fewer collisions

  d As the concentration is halved the particles are less crowded; (as there are fewer particles in the same area) fewer collisions take place

2 a 63 cm³

  b The line levels out (no more gas is being produced)

  c Graph is less steep initially; but ends at the same volume of gas given off

Page 208 Rate of reaction (3)

1 a A large volume of gas is released

  b The fine powder has a very large surface area; which provides a very large surface for contact with oxygen; which can cause an explosive reaction

2 a 0.05 g

  b $CaCO_3$

  c After 6 minutes

  d

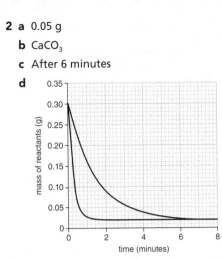

3 a A catalyst is a substance which changes the rate of a reaction; catalysts remain unchanged at the end of the reaction; only a very small amount of catalyst is needed

  b i With copper at 20 seconds and with no substance at 40 seconds

  ii The reaction with the copper went at a higher rate; than the reaction with no substance

Page 209 Reacting masses

1 a 3

  b 5

  c 101 [39 + 14 + (16 × 3)]

  d 5

  e 74   [40 + (17 × 2)]

2 a Mass is neither created or destroyed

  b (Formula) mass of the reactants is 103; (formula) mass of the products is 103

3 a 50 – 28 = 22 g

  b i 40 g

  ii 4.8 g (four times as much)

  c She will need to work out how much zinc there is in the mass of zinc chloride that she needs; use excess acid; and make sure all the zinc is used up

  d $ZnCO_3 \rightarrow ZnO + CO_2$

  (1 mark for all 3 symbols, 1 mark for balancing the equation)

Page 210 Percentage yield and atom economy

1 a Loss through filtration; loss in transferring liquids; loss in evaporation; not all reactants will react to make products   *(Any 3)*

  b i 28 g

  ii 42 g

  c i Percentage yield = $\dfrac{\text{actual yield}}{\text{predicted yield}} \times 100$

  ii 66.7 %   [Percentage yield = $\frac{28}{42} \times 100$]

2 a i How much of the reactants become the wanted products

  ii Not all the reactants became products

  iii So there is no unwanted waste

  b i b

  ii a

c i The M_r of $NaNO_3$ is 85 [23 + 14 + (3 × 16)]; the M_r of H_2O is 18

  ii Atom economy = $\dfrac{M_r \text{ of desired products}}{\text{sum of } M_r \text{ of all products}} \times 100$; $\frac{85}{103} \times 100$; = 82.5%

Page 211 Energy

1 a A reaction which transfers heat to the surroundings

  b A reaction which transfers in heat from the surroundings

  c Measure the change in temperature

  d i Endothermic

  ii Exothermic

2 a B, highest temperature rise

  b Use a spirit burner; heat 100 g of water in a copper calorimeter; measure the mass of fuel at start and after 1 g of fuel is burnt; measure the change in temperature of the water; make sure that both tests are fair. For example, the distance between flame and the calorimeter should be the same; make sure that the tests are reliable. For example, repeat the test three times

  (Any 5)

  c Energy = 100 × 4.2 × 30 = 12 600 J

  (1 mark for mass of 100, 1 mark temperature of 30, 1 mark for correct answer)

Page 212 Batch or continuous?

1 a A continuous process is always in use 24 hours a day; a batch process is started, run for a length of time then stopped and the vessel cleaned out

  b Pharmaceuticals need sterile environments for manufacture; the vessels can be cleaned out more times in a batch process; they are often only needed on a smaller scale as the raw materials are more difficult to source/cost more

  (Any 2)

  c Continuous processes can produce large quantities of products/more product made as the process can keep going day and night; the processes can be automated; there are few labour costs; raw materials can be bought in bulk so costs are reduced

  (Any 2)

2 a Labour; energy; plant costs; raw materials; research and development; safety testing

  (Any 4)

  b It costs a great deal to put in the research and development needed, it can take years to develop; countries have strict safety laws for testing a new medicine; the raw materials needed may be rare and expensive; compounds extracted from plants are difficult to find; plants contain thousands of similar chemicals, so separating the desired chemical is time consuming and expensive/raw materials are sometimes difficult to extract

  (Any 3)

3 a Their safety can be known, without side effects from impurities

  b The plant is crushed to break the strong cell walls and boiled in a suitable solvent so that the chemicals can dissolve; different solvents are used to separate compounds; chromatography is used to separate and identify the compounds; once separated, chemical Z is then compared to P and R by looking at melting points, boiling points and chromatography

  c The data for chromatographic movement is the same for Z and P (2 cm in 10 mins); the melting point is slightly lower than the pure sample P; the boiling point is slightly higher than the pure sample P

Page 213 Allotropes of carbon and nanochemistry

1 a They are all made only of carbon but their atoms are arranged in different ways

  b Diamond; buckminsterfullerene; graphite

  c Allotropes are elements that have the same atoms that are arranged in different ways (these are allotropes of carbon)

2 a They can be combined with graphite to give strength with flexibility

  b They can act as semiconductors

  c They act like a cage to carry the drug and are so small they can travel around the body

3 a Lustrous (shiny); very hard; colourless; transparent; does not conduct electricity; very high melting point; insoluble in water

(Any 4)

  b Opaque; lustrous; soft; slippery; conducts electricity; black/grey; very high melting point; insoluble in water

(Any 4)

  c Diamond is hard; cut diamond reflects light from different angles

  d Graphite is slippery and so when put on paper, some of the surface slides off and leaves a black mark; because it is slippery, it is also a good lubricant

  e In both diamond and graphite, every carbon atom is joined by strong covalent bonds; which form in different directions

Page 214 C3 Extended response question

5–6 marks
Comprehensive descriptions of each experiment, which must include a diagram of the apparatus for measuring temperature change in water over a burning flame and the measurements to be made (mass of water, temperature change, mass of fuel used). If one other factor controlled is included, this may count as a bonus against an error made. A complete explanation of formulae used;

$E = m \times c \times \Delta T$ and E per gram = $\dfrac{E}{\text{mass of fuel burnt J/g}}$

All information is relevant, clear, organised and presented in a structured and coherent format. Specialist terms are used appropriately. Few, if any, errors in grammar, punctuation and spelling.

3–4 marks
The description of the experiment must be correct with a diagram included. Some attempt at the calculation should be made, but the mass chosen may be incorrect.
The information should generally be relevant and presented in a structured and coherent format. Specialist terms should mostly be used appropriately. There may be occasional errors in grammar, punctuation and spelling.

1–2 marks
An attempt must be made at a correct description of the experiment or the use of a diagram as explanation. Answer may be simplistic. There will be limited use of specialist terms. Errors of grammar, punctuation and spelling are likely to prevent effective communication of the science.

0 marks
Insufficient or irrelevant science such as repeating the question. Answer not worthy of credit.

C4 The periodic table

Page 215 Atomic structure

1 a The symbol Co has a lower case letter so denotes an element. CO has two upper case letters, C for carbon and O for oxygen so is a compound

  b Electrons; positive; electrons (have charges that are) negative; neutral; small

  c

	Relative charge	Relative mass
electron	−1	0.0005 (zero)
proton	+1	1
neutron	0	1

(1 mark for each column)

2 a The number of protons in an atom

  b The number of protons and neutrons in an atom

  c

Atomic number	Mass number	Number of protons	Number of electrons	Number of neutrons
19	39	19	19	20

3 a An element that has atoms with different numbers of neutrons/same atomic number but different mass number

  b

Isotope	Electrons	Protons	Neutrons
$^{12}_{6}C$	6	6	6
$^{14}_{6}C$	6	6	8

(1 mark for each row)

4 a 3

  b They are arranged in order of increasing atomic number; in groups numbered with the number of electrons in the outside shell; in periods in order of how many shells the electrons occupy

  c

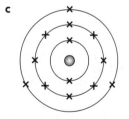

Three shells are needed as there are 13 electrons; that have a pattern 2.8.3 occupying three shells

  d Chlorine

Page 216 Ionic bonding

1 a A charged atom or group of atoms

b
Atom	Ion	Molecule
H	H^+	H_2O
Mg	O^{2-}	CO_2
Cl	Cl^-	Cl_2
Na^+	Br_2	
SO_4^{2-}	F_2	

(1 mark for each column)

c An atom that has extra electrons in its outer shell and needs to **lose** them to be stable. **M**; An atom that has 'spaces' in its outer shell and needs to **gain** them to be stable. **N**

d  LiF

e Atoms are stable when they have a full outer shell of electrons. Lithium has one electron in its outer shell so is more stable if this electron is lost. This leaves the atom with a positive charge

f Fluorine has an electron pattern of 2.7 so needs one more electron to become stable. This becomes a negative ion

g The positive ions of lithium are attracted to the fluoride negative ions; to make a giant ionic lattice which is solid

h $CaCl_2$

2 a Solid X → liquid √ → solution √

b Magnesium oxide

c Giant ionic lattice

d When it is molten or in solution

Page 217 The periodic table and covalent bonding

1 a Ionic

b No

c Electron pairs are shared

d Simple molecules with weak intermolecular force

2 a A group is the elements in a vertical column in the periodic table; O/S/Se/Te/Po
(1 mark for column description, 1 mark for any 3 elements)

b A period is the elements in a horizontal row in the periodic table; Li/Be/B/C/N/O/F/Ne
(1 mark for row description, 1 mark for any 3 elements)

3 a Magnesium is in group 2 as it has two electrons in its outer shell

b Both chlorine and fluorine have the same number of electrons (seven) in their outer shell

c Neon has an electron configuration of 2.8 so is in period 2 as it has two electron shells. Potassium has an electron configuration of 2.8.8.1 so is in period 4 as it has four electron shells

d Group 3

e Period 3

4 a Dobereiner; he saw patterns in the behaviour of groups of three elements

b The behaviour changed periodically/behaviour changed so that every eighth element was similar

Page 218 The group 1 elements

1 a Lithium/sodium/potassium/rubidium/caesium/francium
(Any 3)

b They react with air/oxygen and they react with water

c There will be a very quick reaction; they will float; hydrogen will be given off; an alkaline solution will form
(Any 4)

d When they react with water they form an alkaline solution

e Potassium + water → potassium hydroxide + hydrogen

f There will be a very vigorous reaction; hydrogen will be given off; an alkaline solution will form/rubidium hydroxide will form

g Caesium; it is lower down the group/it loses its outer electron more easily

2 They each have one electron in their outer shell/they each lose one electron to form an ion

3 a Yellow; lilac

b Lithium

c Dip flame-test wire into dilute HCl; dip rod into the group 1 compound and hold in flame; if flame turns yellow/red/lilac it is a compound of sodium/lithium/potassium; other two flame colours matched with remaining two of sodium/lithium/potassium

Page 219 The group 7 elements

1 a Halogens

b Fluorine/astatine

c Sterilising water; to make plastics/pesticides

d To sterilise wounds

e
	Physical appearance
chlorine	green gas
bromine	orange liquid
iodine	grey solid

(2 marks – 3 correct, 1 mark – 2 correct)

2 a Lithium + iodine → lithium iodide

b Sodium + bromine → sodium bromide

c $2K + Cl_2 → 2KCl$
(1 mark for all three symbols, 1 mark for balancing the equation)

3 a Are less reactive going down the group/are more reactive going up the group

b Sodium iodide + bromine → sodium bromide + iodine

c A red–brown solution

d Displacement

e Potassium iodide + chlorine → potassium chloride + iodine

f $2KI + Cl_2 → 2KCl + I_2$
(1 mark for all four symbols, 1 mark for balancing the equation)

P3 Answers

Page 220 Transition elements

1 a Vanadium

b Cu; Fe; V

c copper compounds — often pale green
iron(II) compounds — often orange/brown
iron(III) compounds — often blue

d A substance that changes the rate of reaction but remains unchanged at the end of the reaction

e Iron (Haber process); nickel (hardening margarine)

2 a The breaking down of a compound into two or more substances by heat

b i Zinc carbonate → zinc oxide + carbon dioxide

ii Pass into limewater which turns milky

c Manganese oxide and carbon dioxide

d Copper carbonate → copper oxide + carbon dioxide

e Blue/green solid becomes a different coloured powder (black)

3 a Precipitate

b Add sodium hydroxide solution to three samples; observe gelatinous precipitates; Cu^{2+} ions make a blue precipitate; Fe^{2+} ions make a pale green precipitate; Fe^{3+} ions make an orange/brown (gelatinous) precipitate (accept solid)

Page 221 Metal structure and properties

1 a Lustrous; strong; good conductors of heat

b i Silver and copper both have higher electrical conductivity

ii Aluminium. It has a lower density than the other metals so the plane would fly more easily

iii Lower melting point

2 a The particles of the metals are held together by (strong) metallic bonds

b The ions of the metals are held together by strong metallic bonding making them hard to separate, so a lot of energy is needed to do this

3 a Under very low temperatures

b A superconductor is a material that conducts electricity with little or no resistance

c The potential benefits are loss-free power transmission; super-fast electronic circuits; powerful electromagnets

Page 222 Purifying and testing water

1 a As a raw material/coolant

b Rivers; lakes; aquifers

c i D

ii 320 units

d Dissolved salts and minerals; microbes; insoluble materials

e Sedimentation, filtration, chlorination; chemicals are added to make solid particles and bacteria settle out; a layer of sand on gravel filters out the remaining fine particles, some types of sand filter also remove microbes; chlorine is added to kill microbes

f Older houses sometimes have lead pipes that dissolve slowly into the water; nitrates from fertilisers get in through run-off; pesticides from spraying near rivers *(Any 2)*

2 a Silver nitrate + potassium iodide → potassium nitrate + silver iodide

b White; cream; yellow

c A sodium chloride; B sodium iodide; C sodium sulfate

d Magnesium sulfate + barium chloride → magnesium chloride + barium sulfate

e Precipitation

f Sodium bromide + silver nitrate → sodium nitrate + silver bromide

Page 223 C4 Extended response question

5–6 marks

A comprehensive description of the experiments using flame tests and precipitate tests with sodium hydroxide. Clear observations about the identity of each ion.

Experiments

Flame test needed.

Precipitate reaction with sodium hydroxide needed.

Use of hydrochloric acid and flame-test wire for flame tests.

Observe colour of flame/observe colour of precipitate.

Add sodium hydroxide to solution of the ion (with care).

Identification

Flame colours: lithium – red, sodium – yellow, potassium – lilac.

Precipitate colours, Cu^{2+} – blue, Fe^{2+} – grey/green, Fe^{3+} – orange/brown.

All information in the answer should be relevant, clear, organised and presented in a structured and coherent format. Specialist terms should be used appropriately. There should be few, if any, errors in grammar, punctuation and spelling.

3–4 marks

A correct description of at least one experiment. Accurate identification of the salts tested using this experiment.

The information should generally be relevant and presented in a structured and coherent format. Specialist terms should mostly be used appropriately. There may be occasional errors in grammar, punctuation and spelling.

1–2 marks

An attempt should have been made at the correct description of at least one of the experiments, or an attempt at the outcome of the results for one of the experiments. Answers may be simplistic. There will be limited use of specialist terms. Errors of grammar, punctuation and spelling are likely to prevent effective communication of the science.

0 marks

Insufficient or irrelevant science such as repeating the question. Answer not worthy of credit.

P3 Forces for transport

Page 224 Speed

1 a i Eve

ii Eve covered the same distance as Melissa but in a shorter time

b Distance = 8 x 1.5 = 12 m; speed = 12/0.5; speed = 24 m/s

P3 Answers

2 a Average speed = $\frac{390}{3}$; = 130 km/h

b Car cannot maintain the same speed throughout

c Yes because if 130 km/h is average he almost certainly must have gone faster at some time

3 a A to B/E to F

b 80 s

c 80 m

d D to E

Page 225 Changing speed

1 a Increased uniformly/at 0.33 m/s² from rest; reaching 10 m/s; after 30 s; then constant speed of 10 m/s

b
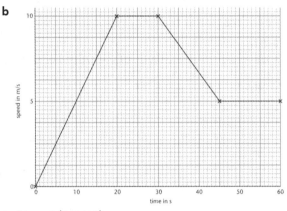

c Area under graph

2 a
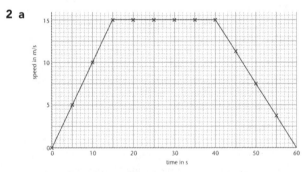

Correct plotting of at least 11 points; uniform increase in speed; constant speed; uniform decrease in speed

b Change in speed = 15 – 3 = 12 m/s; deceleration = $\frac{12}{15}$; = 0.8 m/s²

Page 226 Forces and motion

1 a Accelerate

b i Increase

ii Acceleration increases

c Smaller acceleration

2 a Acceleration = $\frac{40}{20}$; = 2 m/s²

b Force = 500 × 2; = 1000 N

3 a Distance travelled by car between applying the brakes and stopping

b Distance travelled by car between seeing the need to brake and applying the brakes

c Thinking distance plus braking distance

d i Increases

ii Reaction time unchanged so travels greater distance in that time

e Tired; age; under influence of drugs/alcohol; distracted/lacking concentration

(Any 2)

Page 227 Work and power

1 a She is lifting a greater load/she needs to use more force

b She is moving it a smaller distance

c Zero

d Work done = 80 × 2; = 160 J

2 a Same amount of work

b Chris is more powerful than Abi

c Weight = 10 × 60; 600 N

d Work done = 600 × 3; = 1800 J

e Power = $\frac{1800}{8}$; = 225 W

f Power = $\frac{1800}{12}$; 150 W

3 a A

b C

c Fuel pollutes the environment; car exhaust fumes are harmful; carbon dioxide is a greenhouse gas; carbon dioxide contributes to climate change

(Any 3)

Page 228 Energy on the move

1 a Energy of a moving object

b i B

ii C

2 a 1.6 litre diesel

b i Fewer road junctions; fewer speed changes; fewer gear changes

(Any 2)

ii Fuel used = $\frac{96}{24}$; 4

iii Megane has smaller engine capacity

3 a Recharging requires electricity from power stations which do cause pollution

b Advantage – no pollution; no batteries; does not need energy from power station

(Any 1)

Disadvantage – not always sunny; not a constant energy source *(Any 1)*

Page 229 Crumple zones

1 a i Crumple zone

ii Air bag

iii Seat belt

b Seat belt – stretches so that kinetic energy transferred into elastic potential; crumple zone – absorbs some of car's KE by changing shape on impact; air bag – absorbs some of person's KE by squashing up around them

c In case the belt fabric has been overstretched

d ABS brakes; traction control; paddle controls; electric windows *(Any 2)*

2 a Force $= \frac{(25 \times 55)}{0.5}$; $= 2750$ N

b Force $= \frac{(25 \times 55)}{0.002}$; $= 687\,500$ N

Page 230 Falling safely

1 Acceleration is independent of mass

2 a Weight acting vertically downwards; air resistance acting vertically upwards

b Weight greater than air resistance

c The faster she falls the greater the air resistance

d i Terminal speed

ii Balanced/equal in size but opposite in direction

3 a Reduce drag/air resistance; allow them to move quicker

b Crouch over handlebars; tight fitting clothes; shaped helmet

(Any 2)

Page 231 The energy of games and theme rides

1 a GPE; GPE; KE; GPE

2 a B

b Increases

c C

d GPE to KE

e Energy is transferred as sound/heat/friction

f Increase height of B to increase GPE; GPE transferred to KE as the carriage falls; more KE means faster speed

Page 232 P3 Extended response question

5–6 marks

A detailed description of how seatbelts, crumple zones and air bags reduce injuries in a crash situation. Mention is made of the change of shape; absorption of energy and reduction of momentum to zero more slowly. The rate of reduction of momentum is linked to a reduction in force on the body.

All information in answer is relevant, clear, organised and presented in a structured and coherent format. Specialist terms are used appropriately. Few, if any, errors in grammar, punctuation and spelling.

3–4 marks

A limited description of how one or two of the design features reduce injuries in a crash situation. Mention is made of a relevant scientific principle but the description of its application is not detailed.

For the most part the information is relevant and presented in a structured and coherent format. Specialist terms are used for the most part appropriately. There are occasional errors in grammar, punctuation and spelling.

1–2 marks

An incomplete description of how one or two of the design features reduce injuries in a crash situation. There is little, if any, mention of the scientific principles. Answer may be simplistic. There may be limited use of specialist terms. Errors of grammar, punctuation and spelling prevent communication of the science.

0 marks

Insufficient or irrelevant science. Answer not worthy of credit.

P4 Radiation for life

Page 233 Sparks

1 a Polythene will become negatively charged

b i Acetate/perspex

ii Attract them

c Copper is a conductor so the charges will move through it; charges will not stay in one place

2 a So that charge will not build up on her body and affect the components

b i This gets rid of any static charge

ii This will limit the current flowing through Sally

3 a You become charged up by the friction between you and the insulator (nylon); when you touch the metal tap it discharges

b Some clothing gets charged positively and some negatively when it rubs together; opposite charges attract

c The screen of a television becomes charged during operation and will attract small light particles of dust

Page 234 Uses of electrostatics

1 a To remove particles/dust from the smoke

b Fossil fuel power stations

c Positive

d Negative

2 a It becomes charged with the same charge and repels

b Make the coffee jar charged opposite to the coffee granules

(2 marks)

3 a Defibrillation is used to restart the heart/restore regular rhythm

b The shock makes the heart muscle contract

c Hair is a poor conductor; water may conduct the charge across the chest away from where it is needed

d To avoid any other people from receiving a shock

e It is safer as it is only for an extremely short time

Page 235 Safe electricals

1 a i There is a gap in the circuit/not a complete circuit

ii Circuit joined to make complete loop

b i The brightness reduces

ii Ammeter in series; voltmeter added in parallel across the bulb

iii $R = \frac{V}{I}$ or $\frac{6}{0.25}$; $= 24\ \Omega$

2 a i To turn off the current if there is a fault/if he cuts the wire by accident

ii Circuit breaker can be reset/ fuses melt so have to be replaced

b Battery is DC/mains is AC; mains is high voltage/ battery is low voltage

3 a i The live wire carries a high voltage around the house; the neutral wire completes the circuit, providing a return path for the current; the earth wire is connected to the case of an appliance to prevent it becoming live

ii Earth

iii So that even people who are colour blind can identify it easily

(2 marks)

b i Fuses are designed to melt when too much current passes but an iron nail will not melt and will therefore allow too much current to pass

ii The live wire is the one at high voltage; so the fuse is placed in the live wire to break the circuit as close to this as possible

iii If a fault develops where the live wire touches the case; the case becomes 'live'; a large current flows in the earth and live wires; and the fuse blows/melts breaking the circuit

Page 236 Ultrasound

1 a There can be no high and low pressure areas/ it is a longitudinal wave

b $\frac{6\text{ mm}}{4}$ = 1.5 mm

c Ultrasound travels faster in the water than in air

(2 marks)

d The vibrations

e Ultrasound is above the range of human hearing

f In a longitudinal wave the vibrations of the particles are parallel to the direction of the wave; in a transverse wave the vibrations of the particles are at right angles to the direction of the wave

2 a Ultrasound vibrations; pass into the body to the stones; the vibrations break up the stones

b High-powered ultrasound carries more energy; stones need more energy to break them up

3 a To check on the health of the baby

b To investigate heart and liver problems; to look for tumours in the body; to measure the speed of blood flow in vessels when a blockage of a vein or artery is suspected

(Any 2)

c 1 000 000 Hz

Page 237 What is radioactivity?

1 a Average number of nuclei to decay per second

(2 marks)

b i Background = $\frac{750}{30}$; = 25 cps

ii Decrease

iii To take an average; because nuclear decay is random and activity is an averaged out measurement

2 a

type of radiation	charge	what it is	particle or wave
alpha	+2	2 protons + 2 neutrons	particle
beta	−1	electron	particle
gamma	0	short wavelength electromagnetic radiation	wave

(1 mark each correct line or column; 3 marks total)

b I Gamma

ii Alpha

iii Alpha

iv Gamma

v Alpha

3 a The time taken for the activity of a sample to drop to half of the original value/time for half of the original nuclei to decay

b $5\frac{1}{2}$ years; lines drawn on graph to show this

Page 238 Uses of radioisotopes

1 a Radiation always present in the environment

b Because of cosmic rays; they are exposed to more because of the height above the ground

2 Place radioactive tracer in the factory's output; monitor radioactivity down the river

3 a Alpha will be stopped by the smoke particles/beta and gamma would not

b Without smoke, the alpha particles ionise the air; creating a tiny current that can be detected by the circuit in the smoke alarm; with smoke, the alpha particles are partially blocked so there is less ionisation of the air; the resulting change in current is detected and the alarm sounds

4 Carbon-14 is found in all living things; by comparing how much remains in the bones an approximate age can be found

Page 239 Treatment

1 a An x-ray would show if any bones are broken

b Gamma kills bacteria; instruments sterilised as a result

c The x-rays are absorbed differently by materials with different densities; so the plate will show up

d Ben will be alone in the room (the radiographer operates the machine from another small adjacent room)

2 a It can kill cancer cells

b It is placed into a nuclear reactor/made to absorb neutrons

c To protect the staff from gamma radiation

(2 marks)

3 a A radioactive isotope which is introduced to the body; to diagnose a problem

b Tracers will travel around the body to the site of a problem; they can be detected outside the body; this avoids having to cut the patient open

(Any 2)

c Gamma

Page 240 Fission and fusion

1 a From left to right: source of energy first box; steam on/near top pipe; turbine on/near fans; generator on/near far right box; water on bottom of second box

b Source of energy; water; steam; steam; turbines; generator

2 a Nuclear fuel contains enriched uranium; which is radioactive

b A chain reaction out of control is a nuclear bomb

(2 marks)

c i The reactor could overheat and explode

ii Sea water would ruin the reactor; but it had to be used to cool it down

3 a The joining together of two lighter nuclei to make one heavier one

b The experimental results cannot be reproduced by other scientists

c Very high temperatures and pressures needed

(2 marks)

d To share costs/expertise/benefits

Page 241 P4 Extended response question

5–6 marks

A detailed description of how smoke alarms work with and without smoke. Mention is made of the isotope used being an alpha emitter and that the alpha particles ionise the gas molecules in the air causing a tiny current which is detected by the electronic circuit. When smoke particles enter they can block the path of some alpha particles, reducing the amount of ionisation and the current drops. This is detected and sets off the alarm. The reasons why smoke alarms are not a risk to health are discussed, including the fact that alpha particles are unable to pass through the plastic casing of the alarm; smoke alarms are fitted on ceilings some distance away from people and alpha does not travel far in air; a source with a long half-life is used so that the householder does not have to change the smoke detector often.

All the information in the answer is relevant, clear, organised and presented in a structured and coherent format. Specialist terms are used appropriately. There are few, if any, errors in grammar, punctuation and spelling.

3–4 marks

A limited description of how smoke alarms work, lacking in some specific details. Mention is made of alpha particles being given off by the source but an incomplete explanation of why this is not a hazard is given.

For the most part the information is relevant and presented in a structured and coherent format. Specialist terms are used for the most part appropriately. There are occasional errors in grammar, punctuation and spelling.

1–2 marks

An incomplete description of how a smoke detector works. There is little, if any, mention of the reasons why there is no need for concern when using an alpha source in a smoke alarm

Answer may be simplistic. There may be limited use of specialist terms. Errors of grammar, punctuation and spelling prevent communication of the science.

0 marks

Insufficient or irrelevant science. Answer not worthy of credit.